Sustainability and Innovation

Technological, institutional and social innovation promotes economic development and international competitiveness, and can do much to reduce environmental burdens. Thus, innovation is an essential factor in the realisation of the principles of sustainable development. The series 'Sustainability and Innovation' aims to positively contribute to a better understanding of the connection between innovation and sustainability. It is interdisciplinary in coverage, concentrating primarily on the links between environmental and innovation research. Individual titles in the series focus on topical issues and cover both theoretical and empirical aspects: concepts of major practical relevance are also discussed. The series is addressed to researchers and policy makers, as well as to specialists and leaders in business and industry.

Viktor Prokop • Jan Stejskal • Jens Horbach •
Wolfgang Gerstlberger
Editors

Business Models for the Circular Economy

A European Perspective

Springer

Editors
Viktor Prokop
Faculty of Economics and Administration
University of Pardubice
Pardubice, Czech Republic

Jan Stejskal
Faculty of Economics and Administration
University of Pardubice
Pardubice, Czech Republic

Jens Horbach
Faculty of Business
Augsburg University of Applied Sciences
Augsburg, Germany

Wolfgang Gerstlberger
School of Business and Governance
Tallinn University of Technology
Tallinn, Estonia

ISSN 1860-1030 ISSN 2197-926X (electronic)
Sustainability and Innovation
ISBN 978-3-031-08312-9 ISBN 978-3-031-08313-6 (eBook)
https://doi.org/10.1007/978-3-031-08313-6

This Springer imprint is published by the registered company Springer Nature Switzerland AG
The registered company address is: Gewerbestrasse 11, 6330 Cham, Switzerland

Acknowledgement

This monograph was supported by grant No. 20-03037S of the Czech Sciences Foundation.

Contents

Business Models for the Circular Economy: A European Perspective: Aim of the Book and Overview of the Different Chapters

Viktor Prokop, Jan Stejskal, Jens Horbach, and Wolfgang Gerstlberger

1 Introduction and Current State of the Knowledge

Against the background of the current trend in terms of the ongoing growth of production and consumption, it will be challenging to ensure resources for future generations. Increased ecological awareness and the obviously growing scarcity of resources have already led to the introduction of new environmental standards, triggering enterprises, regions, and even countries to adopt new business models (BM) and industrial reconversion approaches. Consequently, many economic actors are seeking to abandon traditional, linear extract–produce–use–dump material and energy flow models (Korhonen et al., 2018) characteristic of the existing (so-called modern) economic systems in favour of models based on the (looped) circular economy (CE) principle, where eco-innovation and sustainable products are created. From the perspective of the EU, the need for academic research to address the environmental challenge, specifically regarding the sustainability and the CE topic, has been declared (Urbinati et al., 2017).

However, despite interest in BM and business model innovation (BMI) increasing dramatically (Velter et al., 2020) and the business perspective offering a more holistic and systemic view in the last 10 years, it still lacks the integration of circularity and sustainability. The traditional BM is directly linked to the economic

V. Prokop · J. Stejskal (✉)
Faculty of Economics and Administration, University of Pardubice, Pardubice, Czech Republic
e-mail: viktor.prokop@upce.cz; jan.stejskal@upce.cz

J. Horbach
Faculty of Business, Augsburg University of Applied Sciences, Augsburg, Germany
e-mail: jens.horbach@hs-augsburg.de

W. Gerstlberger
School of Business and Governance, Tallinn University of Technology, Tallinn, Estonia
e-mail: wolfgang.gerstlberger@taltech.ee

V. Prokop et al. (eds.), *Business Models for the Circular Economy*, Sustainability and Innovation, https://doi.org/10.1007/978-3-031-08313-6_1

value it creates as the main driver is monetary value for shareholders and the company's BM could be defined as a system of interconnected and interdependent activities that determine the way the company "does business with its customers, partners, and vendors" (Amit & Zott, 2012). Therefore, in response to the above, BMI has received increasing attention in specific areas (e.g. the CE, sustainability, servitization, digitization, and social innovation). Nowadays, we can see that different sub-streams have emerged (Pieroni et al., 2019), whereas the first publications of empirical research linking the CE and BM appeared in 2017 and showed a clear increase in 2018 and especially in 2020.

The two sub-streams which have received greater attention in the last 5–10 years, sustainability-oriented and CE-oriented BMI, are the mainstays of this book. The sustainability-oriented BMI sub-stream has evolved significantly over the past decade and incorporates sustainability principles as guidelines for BM design, adding complexity to the conventional BMI process. In addition to generating superior customer value to achieve a competitive advantage and capture economic value, it also seeks to contribute positively to society and the environment. In contrast, the newest concept for the pursuit of global sustainability is currently a CE strategy, although the most important benefit of a more CE-based approach is the possibility of retaining the added value in products for as long as possible, extracting their maximum value and eliminating waste (Smol et al., 2017). More generally, CE-oriented BMI incorporates principles or practices from CE as guidelines for BM design and aims to boost resource efficiency and effectiveness (by narrowing or slowing energy and resource loops). Research on CE-oriented BMI is even more recent than that on sustainability-oriented BMI and has grown rapidly in the last 5 years (e.g. Pieroni et al., 2019; Konietzko et al., 2020).

2 Aim of the Book

Despite the increased interest in BMI and its individual sub-streams, to date, research on these areas has been limited. First, expanding interest in BMI and its sub-streams has led to wide, fragmented, and confused research across various fields, including the innovation management, strategic management, and entrepreneurship literature (Yang et al., 2020). Second, although CE-oriented BM can deliver significant improvements, it also requires managers and practitioners to acquire new knowledge, develop additional skills, and engage in an experimental, iterative, and transformative process with learning cycles and sustainability checks (Chap. 7). However, academia has not yet found consensus on the concept of BMI, its various measures, or what exactly constitutes BMI towards circularity and/or sustainability (Chap. 2). Therefore, to address complex and highly dynamic factors influencing the successful implementation and innovation of sustainability-oriented and CE-oriented BM, it seems necessary to continue this research line.

Concerning the European perspective, there was and still is a need to improve resource efficiency and redefine growth goals with positive social, environmental, and economic benefits. As a result, the issues of sustainability and the shift towards the CE have become cornerstones of European policies and documents (e.g. The EU's Circular Economy Action Plan; the White Paper on the Future of Europe; the European Green Deal), which has also led, among other things, to researchers' increased interest in analysing the issues of sustainability and the CE in the context of European countries. Significant differences have been identified between firms and countries, as evident, for example, between Western and Northern European countries and those in Central and Eastern Europe, where socialist regimes historically did not consider environmental issues to be fundamental or solve them systematically (Chap. 10). The resulting environmental pollution and lower perception of environmental problems in society emphasize the need for a quality, effective, and targeted regulatory function of the state (Horbach, 2016), whereas the countries of Central and Eastern Europe (the so-called Eastern Bloc) have been and are often seen as major European polluters (Tokunaga, 2016). Moreover, research to date has shown that differences also arise between countries belonging to the same group of countries, such as the so-called catching-up countries of Central and Eastern Europe, as shown in the study of Prokop et al. (2022) using an example of neighbouring countries.

Therefore, in response to the above arguments, considering the fact that the issue of pursuing sustainability and orienting more towards CE and other environmentally friendly business practices is on the rise (Chap. 1), *the aim of this book is to provide readers with a closer picture of the issues of business models and their innovation for the circular economy in Europe*. More specifically, this book *aims to provide a unique view of different European perspectives and enable the reader to compare the situation in countries with completely different historical and economic conditions as well as with different perceptions of the need for the actions leading to the sustainable development and to the shift towards the circular economy.*

To achieve these aims, this book is divided into three parts, each comprising three chapters, which allow the reader to look at the issues examined from different perspectives. More concretely, *Part I* provides the reader with new theoretical knowledge but also contains practical examples on the relationship between BM and CE. *Part II* provides the picture from the Southwest European Perspective, and *Part III* is focused on the Central and Northeast European Perspective.

3 Summary of Chapters, New Findings, and Future Directions

3.1 Part I: State of the Art and Major Trends

The book starts with a comprehensive literature review on current research on circular economy business models (CEBM) provided by Hojnik, Biloslavo, and Bertoncel (Chap. 1). The authors explore research gaps, advantages, and disadvantages of the methods used by considering prior research and implications of the research findings for theory and practice. Moreover, the authors conduct bibliometric and topic analysis, which complements and strengthens findings from the systematic literature review. Several trends in the research on CEBMs are identified in Chap. 1, according to de Pádua Pieroni et al. (2018) including:

- *Product-Service Systems (PSS) appear to be conducive to the synergistic application of multiple circular strategies, whereas combining resource efficiency strategies (preventative) with effective resource strategies (to extend products'/ materials' life) seems to be a trend in companies implementing PSS;*
- *circular solutions based on the result-oriented PSS are less frequent, although the potentials for energy savings and resource decoupling are claimed to be higher; and,*
- *companies have room to improve the application of more diversified CE strategies, such as new technology paradigms (refuse), rethinking value propositions' and offerings' fulfilment (rethink), and extending the lifespan of products through refurbishment and remanufacturing.*

Moreover, the authors state that firms with CEBMs are increasingly focusing on design as a first step towards products and clean production processes that bring competitive advantages for partners—not only consumers but also other stakeholders—including the environment. This assertion is linked to the value proposition, waste management, BMI, and product recyclability, which are all focused on creating a closed-loop supply chain in which the products and resources with which they are made are designed for multiple life cycles.

To illustrate differences in current BM and a possible trajectory towards sustainable CEBM, Wecht, Cesinger, Vallaster, and Aleksić (Chap. 2) develop a framework and three design propositions to illustrate archetypes and pathways for how companies can transform their BM from being purely economic driven to a sustainable CEBM. The authors use five examples—COS, Levi's, Patagonia, TOMS, and Zara—representing the apparel and footwear industry. This sector is generally perceived to ignore responsibility for human labour conditions and the environment and to be one of the world's most polluting sectors as well as a significant contributor to climate change and the degradation of the environment. This chapter shows that clear differences exist in how seriously companies take comprehensive sustainability issues, and it proposes that future research should focus on the role of responsible leadership or business purpose, among other ideas, in BM transformation. Furthermore, the authors pose the following questions for future research: *How do companies tackle the transformation of their business models from a traditional business model towards an increased sustainability and circularity? Why do double bottom-line companies possibly experience or opt for mission drift?*

Peralta closes *Part I* with Chap. 3, which is relevant for entrepreneurs, innovators, analysts, governors, and researchers willing to approach lean start-up as a sound practice for achieving sustainable BM that enables a balanced triple bottom line, including economic, social, and environmental goals. More specifically, this chapter provides a comprehensive review of the Customer Development process, suggesting that the lean start-up application in the field of sustainable BMI has been overlooked. The author concentrates on deepening specific lean start-up features, including a set of 17 claims called the Customer Development Manifesto, the core lean start-up process called Customer Development, and the suitability of the Customer Development Manifesto and Customer Development for sustainable BMI. In the final section, Peralta proposes a way to confirm his preliminary validation of the Customer Development methodology using cases, quantitative analyses, and even simulations and encourages a more extensive description and explanation of sustainable BMI challenges and activities from the Customer Development perspective.

3.2 Part II: Southwest European Perspective

The first Southwest European perspective is provided by Carraresi and Bröring (Chap. 5), who consider three cases of new technologies developed in the North-Rhine-Westphalia region of Germany—namely, secondary metabolites' extraction from tomato by-products; phosphate recovery from rapeseed oil press-cakes; and alternative sources of biomass for a new biorefinery concept. The authors outline the challenges associated with the potential implementation of these technological innovations. Results of Chap. 5 show several barriers hampering the valorization of agricultural by-products and the uptake of emerging bio-based technologies. Concerning organizational challenges, the dominance of the existing BM has been revealed as the biggest hurdle in all three case studies provided by the authors. The findings from expert interviews show a widespread hesitation to redesigning current BM in order to implement the novel bio-based process; such hesitation is mainly associated with the uncertain achievable profits and the long time span required to obtain returns. Regulatory and economic aspects are also considered in Chap. 5. From the economic viewpoint, Carraresi and Bröring (Chap. 5) state that, despite the potential advantages of the innovative bio-based processes, some concerns remain associated with the overall profitability of the process and with the quality of the final industrial and consumer products. Chap. 5 also provides information about future research, which, among other goals, aims to extend the sample of companies interviewed to include other potential stakeholders, such as producers' and farmers' associations, consumers, and policy makers. Future research would also need to implement a deeper economic and life cycle analysis to better assess the feasibility of analysed processes as well as their effective environmental and economic sustainability to foster their adoption by the companies involved. In addition, it would be necessary to assess consumers' interest in and acceptance of bio-based products to obtain a more complete overview, including the downstream part of the value chain and the marketability aspects.

In Chap. 6, Andriamanantena, Laperche, and Boutillier highlight the link between the CE and the innovation ecosystem in an industrial port territory of Dunkirk in France and assume that a circular project at the territorial level can lead to the emergence of an innovation ecosystem called "circular innovation ecosystem", characterized by its actors, relationships, and resources and resulting in the creation of value at both the micro- and meso-levels. The authors confirm (a) the characteristics of a circular innovation ecosystem and (b) actors who work in collaboration and share resources to develop circular activities in Dunkirk. In addition, the authors' state that the circular project in this industrial port territory, oriented toward industrial ecology, has been an impetus for the development of such a circular innovation ecosystem, which is still in its development phase. As one of the results, we can see that eco-innovations produced in the territory do not translate to established coordination among the actors. Moreover, despite the actors having the means to develop and carry out eco-innovations, the industrial port territory of Dunkirk remains a territory for the implementation of eco-innovations that are developed outside its borders. From the perspective of regulatory organizations, these organizations too often intervene at the same stages in the development of synergies, which reduces the impact of their actions and can even create confusion as to the role of each of them. In terms of the university, it seems to play a secondary role despite its proficiency and the formations set in place.

Concerning the avenues for future research, to deepen the reasoning behind why such a gap exists between the efforts and results displayed, a study examining the optimization of the role of each institutional actor within the industrial port territory of Dunkirk (circular innovation ecosystem) should be considered. It would also be fruitful to undertake case studies about the way companies transform their BM and identify the barriers in their transition to green BM.

The last chapter in *Part II* (Chap. 7), developed by Kowszyk, Abou-Ali, and Arroyo, deals with the case of Spain. More specifically, the authors focus on the Spanish government's strategic plan for a transition to the CE, called Circular Economy 2030, and its impact on the development of CEBM. Using a comparative analysis of six key sectors and company cases, the authors develop a framework of strategies to help companies transition to a more sustainable and circular BM. Moreover, the authors seek to answer the following questions: *What are the main opportunities or barriers arising, and which firm-level strategies are developed by companies to circularly innovate their business model? How is Spain's Circular Economy 2030 strategy reflected in companies' established ways of doing business?*

The authors also suggest several paths for future research. First, future research should focus on the coordination between business- and policy-driven circular BM development mechanisms and elements. Next, future research could explore how institutions could measure the efficiency of their sustainable policy through circular BMI. Finally, future research could focus on how circular BMI processes influence positive social and societal impacts.

3.3 Part III: Central and Northeast European Perspective

Pizzol and Andersen (Chap. 8) open the final part of the book and provide a detailed and rich picture of the challenges and context of green technology developments within the Nordic countries. The authors investigate how and why the journey of many green technologies, from their early conception to their large-scale adoption and upscaling, remains a serious challenge for the Nordic countries, which are often considered to be leaders in environmental innovation. These countries (a) have rather ambitious policies for environmental protection and low-carbon development and (b) share a long tradition for building a knowledge-based society with substantial public financial support for research and development in general as well as for green technologies in particular (Westholm & Beland Lindahl, 2012; Hildingsson et al., 2018).

Chapter 8 shows that green technology innovation within the Nordic countries generally comprises seven different stages: R&D, start-up, patenting, upscaling, dealing with risk, commercialization, and adoption. Four main factors influence and shape these stages:

- Organizational setting.
- Technology-specific characteristics.
- Funding available.
- Commercial market conditions.

Pizzol and Andersen also provide several challenges for future research, including further analyses of environmental innovations from a gender perspective. The authors suggest that it is worth understanding how ethical concerns (or the lack thereof) are incorporated into financing decisions regarding green technologies from the investor side. They suggest extending the analysis performed in Chap. 8 to other green technologies beyond those included in this chapter and beyond those already extensively addressed in the literature (biofuels, photovoltaics, wind turbines) and to perform comparative analyses of technologies related to the environmental innovation process and innovation in other areas.

Ahmadov, Gerstlberger, and Prause (Chap. 9) aim to help the Baltic region catch up with Western Europe and provide a broad overview of economic instruments by answering the question: *What are the existing fiscal incentives in the Baltic States for the enterprises in their transition towards the circular economy?* More specifically, the authors focus on three former Soviet Republics: Estonia, Latvia, and Lithuania. Concerning their adaptation to European standards, each of these countries faces its own set of challenges regarding the implementation and acceleration of their transition towards a circular economy.

The findings detailed in Chap. 9 confirm the Baltic states' commitment to pursue energy efficiency, research and development, information and communication technologies, and digitalization. Moreover, each state puts efforts into resource efficiency intended to contribute to the development of renewable energy resources as well as the efficient use of resources. The authors state that the Baltic region collaborates with other regions in Europe, especially with the Nordic countries. The BioBaltic project is one example of the Nordic–Baltic cooperation regarding the circular economy.

As this chapter adopted a desk-based research approach, the authors recommend that future research adopt other techniques, such as interviews with state representatives from relevant ministries or investment centres, to delve deeper into this topic through the collection of first-hand data and insights. The authors also propose the need for a deeper understanding regarding fiscal incentives' impact on organizations in relation to the implementation of CE practices. This impact could be studied longitudinally by focusing on the organizations that have received fiscal incentives for the planned activities regarding CE.

Finally, Chap. 10 puts Porter's hypothesis (Porter & Van der Linde, 1995)—which states that pollution could be seen as some kind of resource waste and that a firm's reuse of this waste could lead to the firm's competitiveness, R&D, product innovation, and performance—in the context of the so-called catching-up countries of Central and Eastern Europe (the Czech Republic, Slovakia, Poland, Estonia, Latvia, and Lithuania). These countries are often perceived as having low levels of environmental awareness, making fewer efforts to address environmental issues, and not having a fully mature awareness of the advantages of eco-innovation (Wysocki, 2021). Thus, Prokop, Stejskal, and Nuur aim to answer the following research questions: *Can we confirm the weak and strong versions of the Porter hypothesis in select Central and Eastern European (CEE) countries? Do environmental regulations affect the environmental awareness of firms in the selected CEE countries? How does environmental awareness affect firms' R&D, innovation, and performance in the selected CEE countries?* The authors use a sample of 2139 firms from the six CEE countries studied. The authors confirm the importance of environmental regulations and their influence on firms' R&D but also show that they trigger environmental awareness among firms within the selected CEE countries. However, the authors do not confirm the validity of the strong version of Porter's hypothesis, which expects environmental regulations to impact the performance of firms in the countries studied. Chapter 10 also offers some impetus for future research.

References

Amit, R., & Zott, C. (2012). Creating value through business model innovation: Could your company benefit from a new business model? Consider these six questions. *Strategic Management Review, 53*(3), 40–49.

de Pádua Pieroni, M., Blomsma, F., McAloone, T. C., & Pigosso, D. C. (2018). Enabling circular strategies with different types of product/service-systems. *Procedia CIRP, 73*, 179–184.

Hildingsson, R., Kronsell, A., & Khan, J. (2018). The green state and industrial decarbonisation. *Environmental Politics, 28*(5), 909–928.

Horbach, J. (2016). Empirical determinants of eco-innovation in European countries using the community innovation survey. *Environmental Innovation and Societal Transitions, 19*, 1–14.

Konietzko, J., Bocken, N., & Hultink, E. J. (2020). Circular ecosystem innovation: An initial set of principles. *Journal of Cleaner Production, 253*, 119942.

Korhonen, J., Nuur, C., Feldmann, A., & Birkie, S. E. (2018). Circular economy as an essentially contested concept. *Journal of Cleaner Production, 175*, 544–552.

Pieroni, M. P., McAloone, T. C., & Pigosso, D. C. (2019). Business model innovation for circular economy and sustainability: A review of approaches. *Journal of Cleaner Production, 215*, 198–216.

Porter, M. E., & Van der Linde, C. (1995). Toward a new conception of the environment-competitiveness relationship. *Journal of Economic Perspectives, 9*(4), 97–118.

Prokop, V., Gerstlberger, W., Zapletal, D., & Striteska, M. K. (2022). The double-edged role of firm environmental behaviour in the creation of product innovation in central and eastern European countries. *Journal of Cleaner Production, 331*, 129989.

Smol, M., Kulczycka, J., & Avdiushchenko, A. (2017). Circular economy indicators in relation to eco-innovation in European regions. *Clean Technologies and Environmental Policy, 19*(3), 669–678.

Tokunaga, M. (2016). Regime change and environmental reform: A systematic review of research on central and Eastern Europe. *Economic Review, 67*(4), 339–353.

Urbinati, A., Chiaroni, D., & Chiesa, V. (2017). Towards a new taxonomy of circular economy business models. *Journal of Cleaner Production, 168*, 487–498.

Velter, M. G. E., Bitzer, V., Bocken, N. M. P., & Kemp, R. (2020). Sustainable business model innovation: The role of boundary work for multi-stakeholder alignment. *Journal of Cleaner Production, 247*, 119497.

Westholm, E., & Beland Lindahl, K. (2012). The Nordic welfare model providing energy transition? A political geography approach to the EU RES directive. *Energy Policy, 50*, 328–335.

Wysocki, J. (2021). Innovative green initiatives in the manufacturing SME sector in Poland. *Sustainability, 13*(4), 2386.

Yang, D., Wei, Z., Shi, H., & Zhao, J. (2020). Market orientation, strategic flexibility and business model innovation. *Journal of Business and Industrial Marketing, 35*(4), 771–784.

Part I
State of the Art and Major Trends

Business Models for a Circular Economy: A Literature Review with Bibliometric and Topic Analysis

Jana Hojnik, Roberto Biloslavo, and Tine Bertoncel

Abstract Transitioning to a circular economy (CE) requires product, business model, and ecosystem innovation. This study presents the results of a literature review exploring the state of the art of the research on circular economy business models (CEBM). The chapter explores the research gaps, outlines the advantages and disadvantages of the methods used, and discusses the implications of the research findings for theory and practice. The revisited material consists of 201 articles collected through the Science Direct database. In addition, webscraping and data processing was done on 407 open-source articles on the topic of CEBMs. A data science approach allows for gaining a deeper insight into the research topic by discovering the most important terms, concepts, and findings, together with some bibliometric data, such as the most cited articles, authors, and institutions. Concepts as servitization, product-service systems (PSS), Circular Manufacturing Systems (CMS) approach, service-based business models (SBBM), value proposition, waste management, digital technologies, and product design are the most prominent in the empirical research about CEBMs.

1 Introduction

In the coming decades, an increase in world population is expected and projected to grow to 9 billion by 2050, which poses enormous pressure on limited natural resources while satisfying customer needs and wants. Even the most conservative projections for global economic growth over the next decade suggest that demand for oil, coal, iron ore, and other natural resources will rise by at least a third, with about 90% of that increase coming from growth in emerging markets (Ellen MacArthur Foundation, 2015). Without rethinking how society uses raw material

J. Hojnik (✉) · R. Biloslavo
Faculty of Management, University of Primorska, Koper, Slovenia
e-mail: jana.hojnik@fm-kp.si; roberto.biloslavo@fm-kp.si

T. Bertoncel
Faculty of Organisation Studies in Novo mesto, Novo Mesto, Slovenia

V. Prokop et al. (eds.), *Business Models for the Circular Economy*, Sustainability and Innovation, https://doi.org/10.1007/978-3-031-08313-6_2

in the linear value chain, many of key elements of industry could be exhausted in the upcoming years. Decreasing total natural resource input is critical for sustainability and therefore is mandatory to improve resource efficiency at all levels, from products to industry-wide systems (Kobayashi et al., 2020).

Due to excess resource consumption, a circular economy (CE) aims to return already used resources to the production life cycle economically and ecologically (Klenk et al., 2020). The CE is characterized as an economy that is restorative and regenerative by design and is attracting significant attention from researchers and policymakers alike (Pagoropoulos et al., 2017). The widespread success of the term "circular economy" in the institutional and public debate over sustainability marks important progress. For the first time in history, different and conflicting perspectives on sustainability seem to agree that the economy needs bio-physical inputs (energy and material) for its operation and, therefore, generates outputs in the form of wastes and emissions (Giampietro & Funtowicz, 2020, p. 64). Moreover, the researchers emphasize that the concept of CE implicitly acknowledges the existence of biophysical inputs and outputs as both are circular flows.

Since the issue of pursuing sustainability and orienting more toward CE and other environmentally friendly business practices is on the rise, we aim to review the literature related to scientific research papers focused on CEBMs to gain an understanding of the existing research findings, prominent authors, and key journals and analyze possible open issues relevant to researchers and practitioners. We aim to provide an overview by scrutinizing the available literature through a classic literature review, complemented by a data science approach (webscraping and data processing). The data science approach allows us to provide additional objective support to the findings of the classic literature review by finding commonalities that would likely go unnoticed when manually searching for them in journal publications.

Our main research question is: What are the views, findings, and experiences of academic authors regarding CEBMs, their most prominent elements, and future development?

2 Theoretical Background

The theoretical part is divided into two subsections, namely, circular economy definition and business models and circular economy.

2.1 Circular Economy Definition

The notion of a CE has its roots in industrial ecology originating from the 1970s, specifically, to adopt the concept of resource cycling that exists in the natural environment in industrial systems to improve performance of such systems and

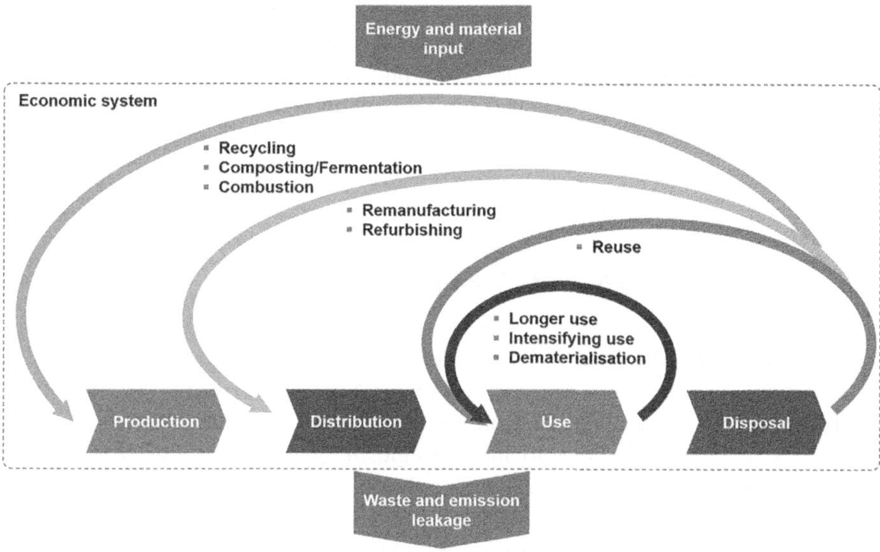

Fig. 1 The circular economy (Geissdoerfer et al., 2020)

reduce the need for the extraction of more resources by closing the product life cycle loop and promoting re-use and recycling of resources (Preston, 2012). The CE is an economic and industrial system in which resources are used for as long as possible – this typically involves businesses implementing a range of alternative business models, such as remanufacture (Muranko et al., 2019). It is a rapidly evolving economic system that benefits society, business, and the environment (Primc et al., 2020). Kirchherr et al. (2017) collected 114 definitions of CE and coded them in 17 dimensions in order to create transparency regarding the current understandings of the CE concept. Their findings indicate that CE is most frequently defined as a combination of reduce, reuse, and recycle activities (likewise as proposed by Geissdoerfer et al., 2020, see Fig. 1); however, the needed systemic shift is not always emphasized. Findings reveal as well that the main aim of the CE is considered to be economic prosperity, followed by environmental quality; its impact on social equity and future generations is barely mentioned. Moreover, neither business models nor consumers are frequently outlined as enablers of the CE (Kirchherr et al., 2017).

The transition to CE implies an innovative approach to production and consumption systems, often requiring radical transformation of business models, related products and services, and supply and value chains to decoupling of resource consumption from value creation (Kravchenko et al., 2020). There are dichotomous approaches to CE, the ones focusing on methods and tools to support industrial practitioners to plan, design, and develop CE solutions, and the others, focusing on methods and tools to predict, monitor, and evaluate CE by measuring its performance and impacts (Kravchenko et al., 2020). In addition, Muranko et al. (2017)

emphasize the need to change business models and, by them, our behavior from linear to circular to conserve energy and resources. According to Muranko et al. (2017), significant influence on the development of CE can be exerted through pro-circular behaviors. Pro-circular behavior is defined as an action that is brought about due to the prioritization of resource efficiency with the aim of supporting the growth of CE.

2.2 *Business Models and Circular Economy*

A business model describes the rationale of how an organization or network of organizations creates, delivers, and captures value (Osterwalder et al., 2010). Transforming a business model from a linear to a CE value chain poses great challenges for companies due to the strategic, long-term decisions that must be taken and that are only reversible at great expense (Klenk et al., 2020). A successful transition to CE requires systemic changes in the way companies understand and do business, with sustainability as a strong foundation (De Pádua Pieroni et al., 2018); therefore, sustainable business model innovation has become fundamental for companies' competitiveness today and even more in the near future. Figure 2 demonstrates the framework for the development of implementation strategies for a global CE. There are two modules. First, the simulation module is where business models are simulated, with a particular focus on the individual behavior of each customer, which is a crucial factor with regard to the success of a CE (Lieder et al., 2017). Second, the optimization module encompasses all aspects of business models (e.g. selection and scaling). The network objects (e.g. opening/closing of locations, change of transport connections) are also considered and analyzed and define the solution space of the optimization itself (Klenk et al., 2020).

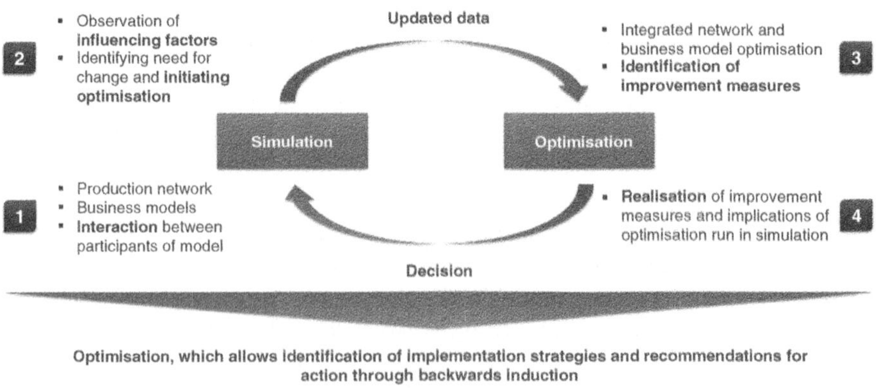

Fig. 2 Framework for developing implementation strategies for circular economy in global production networks (Klenk et al., 2020)

The true success of CE depends on new business models that extract the actual value that is still embedded in products, such as those business models that sell high-quality, long-lasting products, a combination of short-lived and durable products, and services rather than products (Esmaeilian et al., 2018). Different business models offer different opportunities for value creation and resource utilization, as described in the following examples (Esmaeilian et al., 2018):

- Collaborative business models enhance companies' ability to build a partnership.
- Service-based business models increase manufacturers' ability to access and control equipment along its entire life cycle as well as to access new customer segments.
- Cloud-based business models enable businesses to tailor products to individual demands.
- Sharing economy models increase the cost-efficiency of the process and help companies focus on individuals as service providers.
- CE approach enables companies to optimize their value-creation processes.

The rapid development of information and communication technologies (ICT) has enabled the prevailing digital transformation (i.e., digitalization), where physical products can be readily digitized in the virtual space and seamlessly interconnected (Zheng et al., 2019). Researchers (Antikainen et al., 2018) have found that digitalization has the potential to boost the transformation toward a more sustainable CE. It can help close the material loops by providing accurate information on products' availability, location, and condition. Therefore, it also enables more efficient processes in companies, by helping to minimize waste, promoting longer life for products, cutting costs, increasing efficiency of resources, and minimizing transaction costs (Antikainen et al., 2018). Antikainen et al. (2018) have found that digitalization enables novel digital platforms and helps create new kinds of markets based on virtualization of the products and processes – it facilitates easier and more efficient networking, collaboration, and co-creation with stakeholders, including customers. The research findings emphasized that digitalization embraces consumer involvement in the product and service innovation processes while also helping companies reach and interact with consumers better than before (social media is one major enabler in consumer involvement and interactive relationships) (Antikainen et al., 2018).

In addition, digital technologies have enabled the formulation of multiple product-service systems (PSS) with considerable economic, environmental, and societal benefits (Pagoropoulos et al., 2017). Starting from the mid-1990s, eco-efficient product-service systems (PSSs) were indicated in the literature as enablers of a more sustainable and resource-efficient industry through re-use and remanufacturing (Copani & Behnam, 2020). Since the diffusion of re-use and remanufacturing PSS is currently limited and mainly restricted to markets also accepting outdated products (e.g., B2B or emerging countries), product upgrade in remanufacturing was recently introduced. Upgrade cycles allow embedding technological innovation into products while remanufacturing, thus reaching advanced performances and satisfying evolving customers' preferences over time. If coupled with the offering of advanced services, remanufacturing with an upgrade would ease the way to new disruptive PSSs and revolutionize both customers' consumption behavior and the manufacturing business model of companies (Copani & Behnam, 2020). The new remanufacturing with upgrade business models will make remanufacturing one of the main pillars of companies' business value generation, allowing improved management of technology cycles and product obsolescence and consequently maximizing the overall economic and environmental benefits (Copani & Behnam, 2020). Moreover, industries are increasingly adopting service business models (i.e., servitization) to offer physical products and services as a solution bundle to meet individual customer needs (Zheng et al., 2019). The convergence of both digitalization and servitization (i.e., digital servitization) has triggered an emerging IT-driven business paradigm, which was coined the smart product-service systems (Smart PSS) in 2014 (Zheng et al., 2019). As pointed out above, another way to successfully implement CE is by means of product-service systems (hereinafter PSS). Based on the findings of case studies, some trends and points for discussion are outlined as follows (De Pádua Pieroni et al., 2018, p. 182):

- The most recurrent PSS types adopted to enable circular strategies are product renting/sharing, lease, and product-related service.
- Some use-oriented PSS cases presented a limited coverage of circular strategies. However, an opportunity exists to better apply strategies during the PSS operation or use-phase, such as refurbish and remanufacture, or recycling or recovering for the end-of-life.
- Result-oriented PSS is often associated with the highest decoupling potential due to the highest degree of incentives for the provider to implement circular strategies (i.e., all material and energy used to deliver the agreed results become cost factors for the provider).
- The frequency of strategies for smarter product use and operation and manufacture and logistics is low. These strategies are related to preventative actions or efficiency measures to reduce resource consumption from the beginning of life, and they may be essential to start the process with less consumption of virgin resources. Interestingly, some cases presenting the preventative strategies managed to combine them with strategies for extending the product life or closing the loop of materials.

In Fig. 3, authors Widmer et al. (2018) presented a circular PSS supply chain, which combines the principles of CE and PSS supply chain and which is defined as a platform that integrates and utilizes the resources of suppliers and customers to co-create the customized and flexible bundling of products and services throughout its life cycle (He et al., 2016). Also, Witjes and Lozano (2016) pointed out that collaboration between producers and suppliers throughout the procurement process can lead to reductions in raw material utilization and waste generation while promoting the development of new, more sustainable, business models.

In continuation, we demonstrate the selection of the sustainable business model archetypes (Maffei et al., 2019 adapted from Bocken et al., 2014). Archetypes are examples of dynamic-oriented conceptual frameworks or tools for CEBM

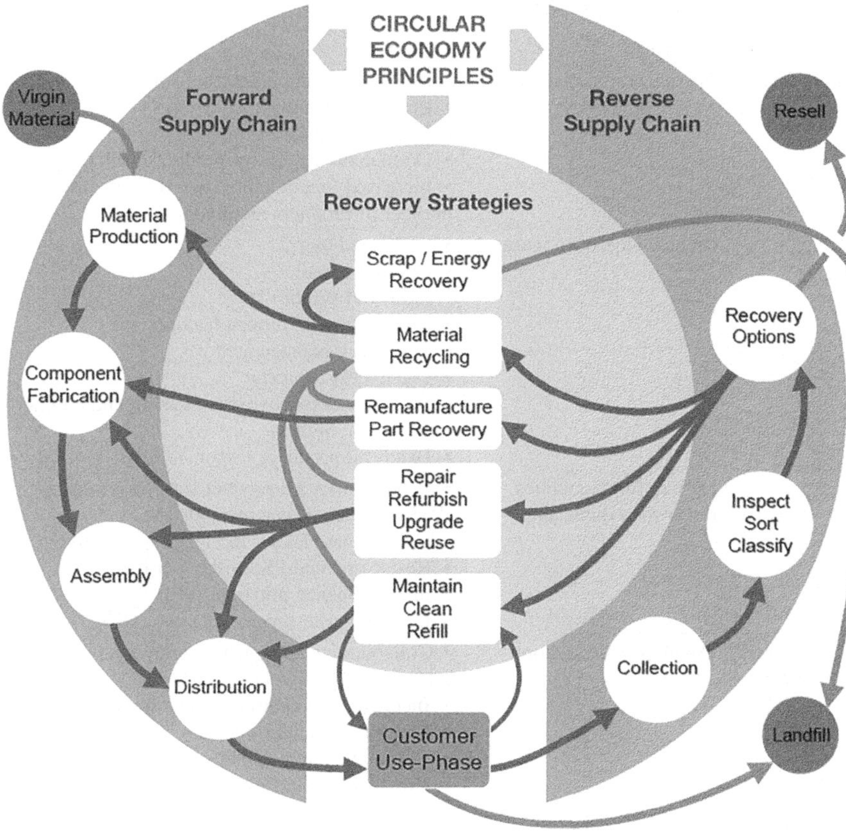

Fig. 3 Circular PSS supply chain (Widmer et al., 2018)

innovation and are meant to describe solutions or configurations of CEBM elements with varying degrees of rigor in the representation of relationships and complexities (Pieroni et al., 2020) (Table 1).

In addition, Pieroni et al. (2020) demonstrate in Fig. 4 how different archetypes are organized according to the source (i.e., downstream or upstream the value system) and type of value generation for CE regarding key elements of a business model (i.e., economic and environmental value captured, value delivered to market or value exchanged with the network/collaborators).

3 Systematic Literature Review: Methodology and Results

Given the research aim of this study, to scrutinize scholarly literature on the CEBMs to identify common elements and patterns as well as to aggregate and evaluate existing insights, we chose a systematic literature review approach since it is

Table 1 Selection of the sustainable business model archetypes

Group	Archetype	Example
Technological	Maximize material and energy efficiency	• Low carbon/manufacturing solution • Lean manufacturing • Additive manufacturing • De-materialization (of products/packaging) • Increased functionality (to reduce the total number of products required)
	Create value from waste	• CE/closed loop • Cradle-to-cradle • Industrial symbiosis • Reuse, recycle, remanufacture • Take-back management • Use excess capacity • Sharing assets (shared ownership and collaborative consumption) • Extended producer responsibility
Social	Deliver functionalities rather than ownership	• Product-oriented product-service system (PSS)—Maintenance, extended warranty • Use-oriented PSS – Rental, lease, shared • Result-oriented PSS – Pay per use • Private finance initiative (PFI) • Design, build, finance, operate (DBFO)
Organizational	Develop scale solutions	• Collaborative approaches (sourcing, production, lobbying) • Incubators and entrepreneur support models • Licensing, franchising • Open innovation (platform) • Crowdsourcing, funding

Source: Maffei et al. (2019)

regarded as the ideal method for this type of research problem (Tranfield et al., 2003). First, the inclusion/exclusion criteria for the systematic literature review and the review protocol were defined (Table 2), as suggested by Jesson et al. (2011).

Sources of data were scientific/research papers published in academic journals and conference proceedings in the period from 2010 until 2020 that included these keywords in the paper's title, abstract, or keywords: "circular economy" and "business model" (see Fig. 5). The period from 2010 until 2020 was chosen considering the fact that the topic of CE in Europe has received enormous attention after the famous sailor Ellen MacArthur made the CE the center of activity of her foundation in 2010 (Stahel, 2020). At the same time, a need for more sustainable business models has also developed, which in the last 5 years is most often connected to the concept of CE (e.g., Diaz Lopez et al., 2019; Pieroni et al., 2019; Konietzko et al., 2020). For the selection of a database, we had three criteria: (a) researchers in the social sciences use this database frequently; (b) the database can be accessed at the library at the authors' university; and (c) a database covers a mix of journals related to the topics of CE and business model. The Science Direct database was thus searched, resulting in 213 papers.

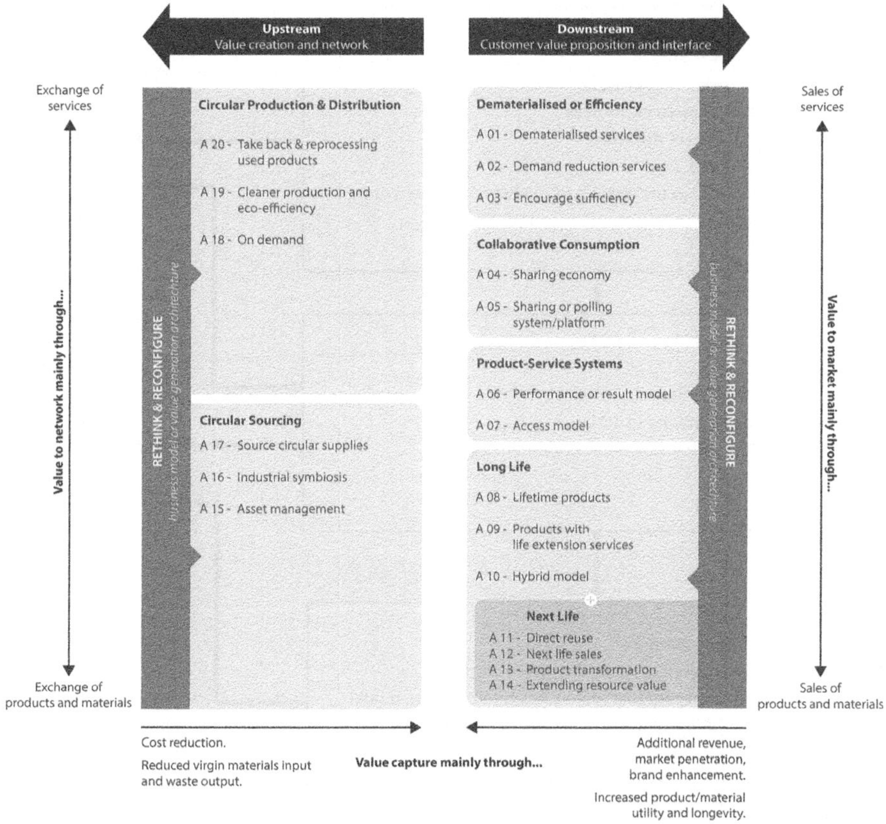

Fig. 4 Consolidated typology of CEBM archetypes (Legend: A archetype) (Pieroni et al., 2020)

Table 2 The selection criteria of the systematic review

Inclusion criteria	Exclusion criteria
• Format: Scientific/research papers • Search keywords included in title, abstract, or keywords • Empirical papers • Period: 2010–2020	• Business model topics with no focus or element of CE • CE topics with no focus on business models • Theoretical papers

In Step 2 we have, among the selected documents, excluded book chapters and brief communication. Not every selected paper addressed the issue of interest. Thus, two authors independently evaluated the 201 papers, and a restricted sample of the 51 relevant papers was obtained and subjected to full analysis as presented in Table 3.

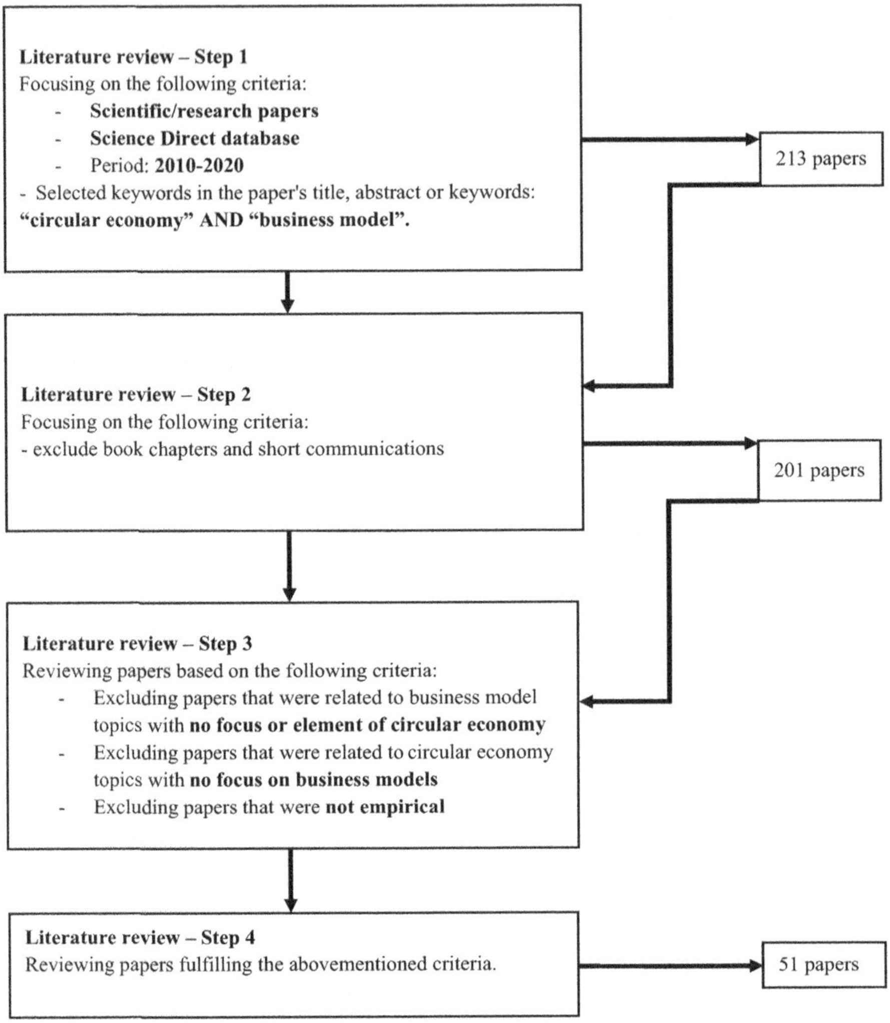

Fig. 5 Collection process of relevant papers

4 Topic Modelling: Methodology and Results

To strengthen the systematic literature review, we applied topic modelling as well, where the webscraping and data processing was done on 407 open-source articles on the topic of CEBMs. In this way, we complemented the database of papers gathered from the Science Direct database and compared our prior findings on the larger sample of papers and with the different method.

The methodology used to study the circular business models included Python libraries for webscraping (urllib, BeautifulSoup, and Selenium), ngram calculation

Table 3 Literature review on circular economy and business models

Authors	Year of publication	Journal	Sample	Key findings
Asif et al.	2018	Procedia CIRP	Case study: Three Dutch companies	"This paper presents a practical ICT framework that is necessary to shift from traditional product sales models to service-based business models. The need for such a framework has emerged from the fact that to achieve the goals of CE, manufacturing industry needs to adopt a Circular Manufacturing Systems (CMS) approach. In this transition, embracing the service-based business models (SBBM) has become inevitable and, for a smoother shift from the conventional sales model to performance-focused SBBM, sophisticated ICT infrastructures are needed. The state-of-the-art review shows that although existing solutions are creating a business case for some small companies, for better management of the value starting from the creation to the recovery and reuse, existing solutions need to go beyond their current functionality. Furthermore, the existing solutions only cover the basic features, which will become outdated with the fast development in the field of IoT" (Asif et al., 2018)

(continued)

Table 3 (continued)

Authors	Year of publication	Journal	Sample	Key findings
Bressanelli et al.	2017	Procedia CIRP	Based on a research project, describes a set of actions and develops very preliminary computations of their expected impact, focusing on the washing machine industry	"Results show that customers could benefit from an average yearly saving of almost 30% of the current washing cost, while country total electricity generation and water consumption could be reduced by about 0.6% and 1% respectively" (Bressanelli et al., 2017)
Bressanelli et al.	2018	Procedia CIRP	Case study of company alpha, a household appliance retailer who operates prevalently in a North European country	"Product-Service Systems (PSS) business models play a major role, as they have been proposed as an opportunity for promoting sustainability and recognized as one of the possible enabling factors to integrate the circular CE paradigm into companies. Nowadays digital technologies are seen as a key factor to enable PSS, allowing at the same time the introduction of CE into companies. This paper explores such a role through the case study of a company which leverages Internet of Things, Big Data and analytics in the provision of its PSS business model. Based on the empirical evidence, the authors confirm the role that the selected digital technologies play in overcoming CE challenges. The Alpha case shows how digital technologies may be used to overcome the CE challenges emerged from literature. More specifically, the case indicates that, following the adoption of IoT, Big Data and analytics, a

				company is able to offer a set of functionalities that, in turn, helps to overcome such challenges" (Bressanelli et al., 2018)
Eguren et al.	2018	Procedia CIRP	890 industrial companies, of which 197 companies have been identified potential to develop their remanufacturing process. They interviewed the industrial managers and directors of innovation of the 18 leading industrial companies in the Basque Country	"According to the interviews, the main motivations that companies have for remanufacturing are strategic advantage and differentiation, high product value and high durability of the product. The main barriers identified are access to products, the resulting quality and market valuation" (Eguren et al., 2018)
Gnoni et al.	2017	Procedia CIRP	A preliminary analysis and the model: Further developments include the model validation and application, to allow a deeper quantitative analysis of the impacts identified and the evaluation of possible strategies to support the implementation of use-oriented business models for CE	"The introduction of new business models, based on use rather than ownership, has been identified as one of the possible enabling actions for the implementation of CE strategies. Thus, Product-Service Systems (PSS) can represent a viable way for companies and customers to switch from a linear to a circular scheme, keeping together the advantages of a customer-oriented offer to those of dematerialization. In this work, an example of an innovative, circular business model for the large appliances sector is proposed, based on a PSS and a closed-loop supply chain. A context study, supported by a deep literature analysis, is performed to identify the main changes involved in the transition from a traditional to a circular supply chain in the sector, as well as the main impacts on the actors involved, through causal loop diagrams. A new circular business model based on leasing and

(continued)

Table 3 (continued)

Authors	Year of publication	Journal	Sample	Key findings
				product recovery and recycling has been described" (Gnoni et al., 2017)
Haanstra et al.	2017	Procedia CIRP	Case study based on two companies. The first company develops and manufactures domestic boilers, primarily for the West European market. The other company is a local recycler situated in proximity to the boiler manufacturer	"This paper proposes a framework to support companies in transitioning from business-oriented sub-optimization to collaborative value chain optimization by visualizing the different relations in a morphological matrix to encourage the selection of the most appropriate principles for the specific industry context. The framework is evaluated in an industrial case with multiple stakeholders, resulting in a feasible closed material loop. Combining the theory of General Morphological Analysis with Life Cycle Engineering resulted in a concise framework for practical application in industry" (Haanstra et al., 2017)
Moreno et al.	2017	Procedia CIRP	ShoeLab case study. ShoeLab is an initiative collaboration by Cranfield University, Cisco Systems, and The Clearing—a branding consultancy, which aims to develop a proof of principle for a smart and sustainable shoe	"This paper examines how the use of Re-Distributed Manufacturing (RdM) and a Product Service-System (PSS) approach, while leveraging the latest digital manufacturing technologies, enables the shift to this new economic model. The ShoeLab case study includes an outline of the business model options supporting this new approach to sustainable production, highlighting the circularity that may be achieved in employing RdM and the latest digital manufacturing technologies in the

				form of 3-D printing. The research conducted indicates that using IDEF0 modelling could help to realize the full potential of RdM such as the manufacturing and transport of products involving less material, energy and waste" (Moreno et al., 2017)
Panarotto et al.	2017	Procedia CIRP	The field study was done with a road compaction equipment manufacturer. Empirical data was collected during meetings (held biweekly)	"The PSS context is relevant as the company is exploring new PSS-like business models in conjunction with a new concept for an asphalt compactor which preliminarily shows the potential to promote a fully circular business scenario. Successful consideration of CE aspects in PSS design requires the ability to make thoughtful trade-offs between technical aspects (costs, performances), PSS business strategy, and the benefits of a circular design, which may manifest years ahead" (Panarotto et al., 2017)
Pieroni et al.	2018	Procedia CIRP	9 case companies that have adopted CE principles in the Nordic region were identified through official reports funded by the Nordic Council of Ministers and in databases of industry associations (e.g. Technology Industries of Finland, innovation Center Iceland)	"This paper explored how different PSS types can enable a number of circular strategies for resource productivity. The main trends are: – PSS appears to be conducive to the synergistic application of multiple circular strategies; – Combining resource efficiency strategies (preventative) with resource effective strategies (for extending products/materials life) seems to be a trend in companies implementing PSS;

(continued)

Table 3 (continued)

Authors	Year of publication	Journal	Sample	Key findings
				– Circular solutions based on the result-oriented PSS are less frequent, although the potentials for energy savings and resource decoupling are claimed to be higher; – Companies have room to improve the application of more diversified CE strategies such as new technology paradigms (refuse), rethinking value propositions and offerings' fulfillment (rethink), and extending lifespan of products through refurbishment and remanufacturing" (Pieroni et al., 2018)
Sousa-Zomer et al.	2017	Procedia CIRP	A larger manufacturer of household appliances that implemented a new circular business strategy as part of its business	"The company's various environmental initiatives target different parts of a product's life cycle, from design to the product's end-of-life phase. All activities performed across the PSS life cycle are oriented and valuable to extending the product life and reducing the consumption of resources, in alignment with CE requirements. Cost savings can be obtained through the activities carried out across the life cycle stages, although this aspect was not yet assessed, since this work is still in progress. Design strategies aligned with the business strategy are adopted. A close consumer relationship strategy during the use phase is employed, allowing performance of continuous improvements in the business models.

				Partnerships with stakeholders at the end of life and the adoption of distinct decision paradigms (e.g., remanufacturing or recycling, depending on the product conditions) are also employed, which may also contribute to achieving resource efficiency" (Sousa-Zomer et al., 2017)
Stewart et al.	2018	Procedia CIRP	Case analysis, based on primary (i.e., emails and semi-structured phone interviews) and secondary data (i.e., from actors' website, annual, and sustainability reports)	"The paper shows how a business model can be elaborated to implement the CE strategy of a beverage producer aiming to close the loop of aluminum beverage cans. Both environmental and economic relevance of the business model were emphasized. High upfront investment for non-core activities, the need for alliance with other beverage brands, the need for cooperation within the value chain with reprocessors, can producers and retailers, high dependence on consumer behaviors, and resistance from powerful actors in the current context who would be excluded in the new business model are all challenges for the case company. There is a need to internalize environmental impacts in business costs which could create incentives for value chain actors to improve the system altogether, to better understand what would drive best recycling behaviors among consumers, and to design business model changes which carefully take into account the current situation and power distribution among actors" (Stewart et al., 2018)

(continued)

Table 3 (continued)

Authors	Year of publication	Journal	Sample	Key findings
Uçar et al.	2020	Procedia CIRP	Three case studies based on secondary sources: Alpha case, Philips CityTouch case, and ZenRobotics case	"Based on the findings, the enabler role is more dominant than the trigger role. The trigger role only exists in the value proposition by offering a use-oriented PSS, in key resources by creating intelligent products and in revenue streams by making the pay per use logic possible. It means that this role is about the reuse option and it initiates some elements of this business model type. The functionalities behind this role are data collection, exchange, and storage and data analysis. The detailed functionalities about data analysis are behind monitoring a product and creating an intelligent product according to case findings and they are linked to the Internet of Things. However, the enabler role facilitates eight building blocks of BMC, namely, partners, activities, key resources; value proposition; customer relations, channels, and cost structure and revenue streams" (Uçar et al., 2020)
Todeschini et al.	2017	Business Horizons	Six interviews with specialists in sustainability, business model innovation, and the fashion industry, along with eight case studies on innovative fashion start-ups that researchers define as "born sustainable"	"Researchers found that sustainability-oriented BMI in fashion tends to be markedly different for incumbents and startups. Secondly, regarding the role of technology in enabling sustainable BMI for fashion startups, it seems that high technology is helpful when it comes to radically rethinking manufacturing processes but is not a

				hard requirement for successful innovative fashion business models. Several cases highlighted the importance of partner commitment and engagement. This leads to at least two relevant strategic issues: how to align values and interests along a typically complex value network and how to promote the sustainability aspects of the value proposition (and, indirectly, promote consumer education) when innovative and sustainable business models in the fashion industry are partnering with indirect sales channels. The cases also suggest typical challenges for scaling up fashion startups. Among these challenges, a common theme is that of keeping true to the core sustainability values, which are strongly tied to essential drivers such as fair trade, commitment to recycling, upcycling, and use of sustainable materials. A third insight refers to the critical importance of a supporting business ecosystem of like-minded firms that share sustainability values. Shared values promote knowledge and resource sharing and facilitate the establishment of collaborative efforts that foster business model experimentation, especially at the startup stage" (Todeschini et al., 2017)
Parida et al.	2019	Journal of Business Research	6 interviews within the manufacturing company (ecosystem orchestrator) and interviews within the associated ecosystem partner's organization	"The paper provides evidence that ecosystem orchestrators achieve the transition toward a CE in two stages: (1) ecosystem readiness assessment, and (2) ecosystem

(continued)

Table 3 (continued)

Authors	Year of publication	Journal	Sample	Key findings
				transformation. In each stage, specific and complementary mechanisms are deployed. The article elaborates on ecosystem transformation mechanisms and their purpose, use, and interdependencies in moving toward a CE paradigm" (Parida et al., 2019)
Åkerman, et al.	2020	Geoforum	In-depth case study of one pioneering biogas company in Finland	"The paper investigates the material politics of the CE by means of an in-depth case study of how one pioneering biogas company in Finland struggled to undo a disruptive regional manure surplus by turning it into a resource in energy and fertilizer production. They treat this novel nutrient recycling business model as a political technology that aimed to rearrange existing metabolic relations in order to solve the problem of intensive pig farming that generated too much manure for sustainable utilization in the area. At the same time, their business model clashed with the various categorizations and technologies of governance that stabilized the status quo. These clashes embody the underlying societal tensions of the CE transition, which do not manifest themselves as open conflicts in public arenas" (Åkerman et al., 2020)
Copani and Behnam	2020	CIRP Journal of Manufacturing Science and Technology	19 firms are selected for the case studies, with the "service offering" as the unit of analysis	"Based on case study results, a model was proposed to classify different types of remanufacturing/upgrade PSS business

				models. They differ for three main variables: the inclusion of performance-based services in the value proposition; the systematic approach in implementing remanufacturing with upgrade as a joint technical process; the extent of exploitation of the remanufacturing/upgrade process in the company value proposition. Case studies show that remanufacturing with upgrade is not yet a mature PSS since, according to the sampling criteria, only 19 companies were found to implement remanufacturing and upgrade. Among these, no company is exploiting joint remanufacturing and upgrade as an enabler for the offering of performance-based services. Companies are either embracing a product-related service strategy based on the selling of remanufactured or upgraded products or are offering performance-oriented services leveraging on the increased functionalities of newly built products, thus not recurring to the potential of remanufacturing with upgrade" (Copani & Behnam, 2020)
Khan et al.	2020	CIRP Journal of Manufacturing Science and Technology	Five case studies on midlife upgrades to analyze the similarities and differences between capital equipment replacements and midlife upgrade decisions	"The findings indicate that midlife upgrades have the potential to effectively postpone replacement while extending the remaining useful life of capital equipment and thereby facilitate the implementation of the CE. Moreover, findings also suggest that midlife upgrades indeed present a value-adding alternative to traditional equipment

(continued)

Table 3 (continued)

Authors	Year of publication	Journal	Sample	Key findings
				replacement from the perspectives of both users and manufacturers, especially when provisioned with a servitized PSS business model" (Khan et al., 2020)
D'Amato et al.	2020	Forest Policy and Economics	Interview with managers in 13 Finnish SME companies from the field of packaging, textiles, composite materials, cosmetics, and pharmaceutical products	"The paper outlined the main business model archetypes and identified the key characteristics that enable value capture and delivery for various stakeholders. Based on their findings, researchers provide insights and recommendations for researchers and policymakers to advance the sustainability transition to a circular bioeconomy in the context of the forest-based industry, and for the management of SMEs to reflect on company viability and growth. While the variety of business models observed in this study support sustainability-based value creation, they appeared to be strongly dominated by traditional practices (e.g., use of renewables, efficiency). On the other hand, more avant-gardist perspectives were missing in the business models analyzed, such as promoting frugality, reducing materiality, securing livelihoods and/or supporting natural systems" (D'Amato et al., 2020)
Sarc et al.	2019	Waste Management	394 companies from the waste management and disposal sector of the DACH region (Germany, Austria, and Switzerland)	"In order to be able to successfully implement the technologies, it is necessary that the systems adapted to customers are coordinated with industrial companies (recovery

				industry and their suppliers). On the one hand, the quality requirements of the industry (which, among other things, uses secondary raw materials and energy) are becoming even higher; on the other hand, the requirements are becoming increasingly stringent from the point of view of environmental legislation. In order to deliver the required material qualities, optimizations in logistics and plant concepts are necessary. The methods of digitalization will make it possible to achieve these improvements. In addition, new business models based on digital technologies will increasingly evolve, contributing to the success of digitalization along the value chain. Currently, different digital solutions such as electronic invoices, service portals for customers, paperless order processing, container tracking, vehicle tracking and tracking for dispatchers or level measurement in waste containers are especially in use" (Sarc et al., 2019)
Jabbour et al.	2020	Resources Policy	Two Brazilian traditional manufacturing firms which are intensive in metallic natural resources	"The main contributions of this study can be summarized thus: (i) there is no homogeneity in companies' understanding of the sharing economy's potential as a win-win approach; (ii) transitioning towards the sharing economy relies on radically changing the way firms develop and upgrade their portfolio of products; (iii) transitioning towards the sharing economy is more

(continued)

Table 3 (continued)

Authors	Year of publication	Journal	Sample	Key findings
				complex – due to its challenges, drivers, and barriers – than has been reported by the literature; (iv) more advanced companies, in terms of sharing economy operations management practices, have mastered the capability of combining key principles of the CE and key elements of the sharing economy, creating an important synergy between the CE and the sharing economy; (v) emerging industry 4.0-related technologies seem to play a vital role in unlocking the sharing economy, as applied to product development" (Jabbour et al., 2020a)
Diaz Lopez et al.	2019	Ecological Economics	143 cases of the implementation of various resource efficiency measures (REMs), the related business model changes (BMCs), and implementation barriers (IBs)	"Findings of this study indicate that the Resource Efficiency Measures (REMs) have been related to clear classes or Business Model Changes (BMCs) that can support their implementation, notably changes in the supply chain (SC), internal processes (IP), customer interface (CI), financial model (FM) and the value proposition (VP). The BMCs were further characterized in terms of typical Implementation Barriers (IBs) that were reported in the cases, i.e., institutional, market, organizational, behavioral and technological barriers. The study could not confirm some common theoretical wisdom, such as that firms mainly focus on 'simple' REMs like cleaner production and

				green products. Internal processes BMCs were mainly hampered by institutional and technological factors. Value proposition and financial model BMCs faced mainly behavioral and market barriers. Customer interface BMCs additionally encountered organizational barriers, while supply chain BMCs face a mix of all classes of barriers distinguished in this study" (Diaz Lopez et al., 2019)
Bauwens et al.	2020	Ecological Economics	2 × 2 scenario matrix method developed through a thought experiment and a focus group	"Key drivers of change in this matrix are the nature of technologies deployed – high-tech or low-tech innovations – and the configuration of the governance regime – centralized or decentralized. This paper builds four scenario narratives for the future of a CE: "planned circularity", "bottom-up sufficiency", "circular modernism", and "peer-to-peer circularity". It delineates the core characteristics and the upsides and downsides of each scenario. It shows that a CE can be organized in very contrasting ways" (Bauwens et al., 2020)
Aid et al.	2017	Resources, Conservation & Recycling	6 in-depth interviews performed with high-level representatives of two of the largest public and two of the largest private waste management (WM) organizations in Sweden	"Eco-industrial park management and contracting out holistic resource management are some areas in which the respondents see waste management (WM) organizations offering new services. In relation to such approaches, various risks (e.g., being cut out of investment benefits, or unstable supply) and opportunities (e.g., new markets and enhanced sustainability

(continued)

Table 3 (continued)

Authors	Year of publication	Journal	Sample	Key findings
				profiles) were identified. Additionally, it was seen that WM companies would need to make substantial changes to their business approach, becoming less dependent on flows of mixed materials for example, if they are to become even more central value chain actors. To strengthen such approaches, it was seen that the sector will need to find methods to strategically build strong, long-term partnerships, expand upon and take advantage of available knowledge resources (i.e., best practice technologies and regional material flows), and explore new business models (i.e., stockpiling, park management, or waste minimization)" (Aid et al., 2017)
Ertz et al.	2018	Resources, Conservation & Recycling	The data collected about these organizations is qualitative in nature but is analyzed quantitatively using a classification algorithm and clustering dimensions from the PLEBM (product lifetime extension business model) framework, in order to build the taxonomy inductively from the data	"The article develops and empirically validates a methodology to classify a specific type of circular business model, namely product lifetime extension business models, involving organizations and consumers, to bring quantitative rigor to conduct and presentation of taxonomy research in the field of the CE" (Ertz et al., 2019)
Gall et al.	2020	Resources, Conservation & Recycling	Case of a for-profit company located in Nairobi, Kenya	"Economic incentives, trust building measures, and a general willingness to learn and adapt were identified as prerequisites for establishing accountable supplier-buyer relationships. High-quality mechanical

				recycling of plastic wastes under informal conditions seems feasible and may even come along with socio-economic benefits for marginalized waste pickers when suitable modes of cooperation are put in place" (Gall et al., 2020)
Guzzo et al.	2020	Resources, Conservation & Recycling	9 medical device industry circular business models (CBM) types	"An adapted version of the business model canvas describing each CBM supports dissemination and discussion. While identified CBMs activate the known technical cycles, there is potential to combine different types of circular solutions and enhance triple bottom line benefits" (Guzzo et al., 2020)
Hopkinson et al.	2020	Resources, Conservation & Recycling	3 illustrative examples used within a global executive education program	"The paper explores the role and interplay of four configurable 'building blocks': circular design, business models, reverse network management and system enablers, as a potentially useful heuristic to describe how businesses are realizing value from their CE practices. These cases illustrate that the success of large-scale value creation and capture derives from the iteration of multiple, boundary-spanning activities emerging over time in varying configurations. There is now a need to move from classification and description to quantification and testing of how value is created and captured from CE in different contexts" (Hopkinson et al., 2020)
Ingemarsdotter et al.	2020	Resources, Conservation & Recycling	12 semi-structured interviews with stakeholders from the company	"The identified opportunities for IoT to support circular strategies in the studied case demonstrate that IoT can support a

(continued)

Table 3 (continued)

Authors	Year of publication	Journal	Sample	Key findings
				servitized business model, support tracking and record keeping of in-use and post-use products, enable conditions monitoring and predictive maintenance, improve estimations of remaining lifetime of used products, and inform design decisions to improve durability of products. These findings add IoT-specific insights to previous literature on challenges in circular business model implementation. Researchers also extracted general (i.e., not IoT specific) challenges regarding the financial uncertainty and limited customer acceptance of circular business models" (Ingemarsdotter et al., 2020)
Krystofik et al.	2018	Resources, Conservation & Recycling	Case study of Davies Office, Inc. (Davies) remanufacturing processes for office furniture	"This case study identifies and explores factors of Davies Office, Inc. (Davies) remanufacturing processes for office furniture that affect the economic and environmental practicality of creating multiple remanufacturing cycles. Life Cycle Assessment (LCA) is used to estimate the impacts of multiple remanufacturing cycles and how these are affected by "adaptive remanufacturing," a neologism to describe the use of an end-of-life (EOL) product core to create a similar, but non-identical product. LCA results suggest that adaptive remanufacturing is both an environmentally preferable and economically viable business

				strategy. Specifically, the ability to update, reconfigure, and customize previously obsolete products to meet present market demands enables life cycle extension beyond what is achievable with traditional remanufacturing. The study posits that such adaptive remanufacturing techniques not only expand the potential environmental benefits of remanufacturing but enhances the long-term economic viability of remanufacturing in the durable product market" (Krystofik et al., 2018)
Leder et al.	2020	Resources, Conservation & Recycling	Focus group discussion with academics, practitioners, and policymakers	"Results indicate that collaborative aspects aid in creating value within this specific business model, as well as that technology can act as a facilitator for collaboration between stakeholders in realizing the benefits of the Waste Valorization model" (Leder et al., 2020)
Nußholz et al.	2019	Resources, Conservation & Recycling	Case study of three pioneering Scandinavian companies that produce circular building materials	"The results show clearly that all three cases offer potential for carbon savings. As the savings vary significantly, findings suggest that careful consideration of affected processes and markets is a key to attaining carbon savings. Business model innovations to enable secondary material use involve establishing key partnerships to access secondary materials, developing recovery process and technology, targeting customer segments that value lower environmental impacts, and considering life cycle costs" (Nußholz et al., 2019)

(continued)

Table 3 (continued)

Authors	Year of publication	Journal	Sample	Key findings
Pieroni et al.	2020	Resources, Conservation & Recycling	The application of the archetype cards in six manufacturing companies	"By reviewing academic and practical literature, this article identified and systematized 20 archetypes of BMs fit for CE according to a typology framework based on the source (i.e., downstream or upstream) and type of value generated to the market (i.e., mainly through services or products), exchanged with the network of collaborators/partners (i.e., mainly through services or materials/energy exchange), and captured by the organization/shareholders (i.e., additional revenues or savings). Based on the initial analysis of archetypes and application of the card deck with 6 manufacturing companies, key findings and insights were identified: (a) existing archetypes hold a reductionist approach with limited attention to the downstream value logic; (b) practical application of archetypes requires improvement towards contextualization to sectorial examples and guidelines on how to select archetypes; (c) standardization of the abstraction level and modifications in the structure of archetypes is necessary; and (d) action research facilitates exploration of more radical forms of innovation for CEBMs" (Pieroni et al., 2020)
Saidani et al.	2018	Resources, Conservation & Recycling	Field-based research that includes an immersion on the industrial ground during a	"Key factors, i.e., regulations, business models and markets evolution, new and

			5-month internship at a major NRMM (non-road mobile machinery) manufacturer, surveys, face-to-face meetings, telephone interviews, and participation at workshops and conferences related to the subject	emerging technologies integration, affecting the CE performance of automotive and HDOR (heavy-duty and off-road) sectors are analyzed. Both industries are compared through the four buildings blocks of the CE and the four possible feedback loops defined by the Ellen MacArthur Foundation" (Saidani et al., 2018)
Sousa-Zomer et al.	2018	Resources, Conservation & Recycling	A manufacturing company implementing a circular business model in a developing country (Brazil)	"The main findings reveal that changes needed for implementing a circular business model are achieved by integrating and engaging multiple organizational functions, each of which is complementary and mutually supportive. The involvement and collaboration of multiple stakeholders play an important role in circular business model implementation and operation. Moreover, the implementation of such models should consider specific contextual aspects of the business in question, i.e., alignment between external conditions and the business model should occur" (Sousa-Zomer et al., 2018)
Ünal et al.	2019	Resources, Conservation & Recycling	Case Bark House, a SME from the USA, which is the first and only Cradle-to-Cradle (C2C) Platinum-certified product holder as of 2017	"Among the main results, essential outcomes included configuring and adapting the company's business model to particular internal and external contextual factors; and valorization of local waste by harmonizing managerial practices, and socio-cultural and socio-economic settings, as well as sustainable behaviors among the actors of the supply chain" (Ünal et al., 2019)

(continued)

Table 3 (continued)

Authors	Year of publication	Journal	Sample	Key findings
Whalen et al.	2018	Resources, Conservation & Recycling	Two case studies of Swedish ICT "gap exploiter" companies (third-party firms that create value through the re-utilization of existing products)	"This paper provides an in-depth look at how CE strategies can be embodied in the gap exploiter business model within the ICT sector. The main findings include barriers to business operations and opportunities to address them through business model innovation and policy intervention. The study furthers understanding of ICT reuse operations and their related business models. It also contributes to the upcoming policy debate on how to encourage and create a more resource-efficient and circular economy" (Whalen et al., 2018)
De Angelis, Feola	2020	Journal of Cleaner Production	A single, exploratory, qualitative case study of Italian company Naturalmente Colore	"The ReSOLVE framework can provide a simple yet effective tool to guide the implementation of CE principles in a practical context, amid the growing complexity of CBM taxonomies and typologies. CBM literature suffers from a lack of definitional clarity which hinders consistent, comparable and detailed understanding of its characteristics" (De Angelis & Feola, 2020)
Geissdoerfer et al.	2018	Journal of Cleaner Production	Four case studies – Alpha is an office furniture remanufacturer, Beta is a high recycled content flat aluminum sheet manufacturer, Gamma designs and produces luxurious fashion accessories from fire hoses, and Delta provides a bike-sharing service	"Development of CBMs demands not only technical innovations (e.g. in terms of material flows), but also social innovations (e.g. in terms of changes in consumer behavior). CBM can contribute to sustainable development by promoting economic, environmental and social goals;

				pro-actively managing stakeholders; including a long-term perspective; and closing, slowing, intensifying, narrowing and dematerializing resource loops" (Geissdoerfer et al., 2018)
Guldmann and Huulgaard	2020	Journal of Cleaner Production	12 case companies comprised of circular start-ups and linear incumbent companies of different sizes and from different industries (i.e., the apparel, textiles, furniture, machinery, and mechatronics industry) that served different customer segments (i.e., B2C and B2B companies)	"Most companies experienced barriers to adoption of CBM at all four socio-technical levels, i.e., barriers at the market level, the institutional and value chain level, the organizational level and the employee level. The comparison of companies of the same size, in the same industry and serving the same customer segment showed that these companies experienced four different barriers, i.e., investments in existing manufacturing facilities and value chain; reluctance to involve external stakeholders in CBMI activities; difficulty establishing cross-organizational collaboration; and a hesitant approach to promoting the CE agenda" (Guldmann & Huulgaard, 2020)
Gusmerotti et al.	2019	Journal of Cleaner Production	A total of 821 manufacturing firms in Italy completed the questionnaire sent by email	"CE principles should be integrated in a synergetic way within the business strategy of a company in order for its business model to be effective. This means that "circularity" should pervade the whole business and, therefore, encompass all business functions: from raw material purchasing to product design, operational activities and market communication. A focus limited to internal operations may reduce the potential economic and market opportunities connected

(continued)

Table 3 (continued)

Authors	Year of publication	Journal	Sample	Key findings
				to the CE, while an exaggerated focus on *communication and marketing actions* (not linked to a cleaner and more circular production system) could be interpreted by the market as greenwashing. In order to effectively integrate and reinforce circularity and efficiency/competitiveness, managers should adopt a life cycle perspective capable of embedding the principle of "circularity" in all the stages of the value chain: from product design, through procurement decisions, production processes, distribution and logistics, to the engagement of customers and consumers as partners (and not only purchasers)" (Gusmerotti et al., 2019)
Henry et al.	2020	Journal of Cleaner Production	127 CSUs (circular startups) in the three geographical areas under scrutiny. A total of 68 (54%) companies are located in the Randstad region, 28 (22%) are from Berlin, and 31 (24%) are from London	"Different typologies of CSUs exist: design-based CSUs, adopting circular innovations mostly in the pre-market phase through source material minimization, product design or production process efficiency; waste-based CSUs, seeking to extract value from unexploited external waste streams; platform-based CSUs, pursuing sharing/trading business models built around B2B, B2C or C2C marketplaces; service-based CSUs, embedding products in service-systems to increase usage efficiency; and nature-based CSUs, increasing the delivery of (products and) services based on nature-

				based systemic solutions" (Henry et al., 2020)
Heyes et al.	2018	Journal of Cleaner Production	The BECE framework was applied to a micro-ICT business in the UK to identify and analyze opportunities for building CEBMs	"ICT support companies are well suited to provide ReSolve actions. However, six ReSOLVE actions appear more appropriate than others for service-oriented technology companies. Other actions do not lend themselves easily to services, according to this research, and so may be less frequently pursued by those unwilling to consider innovations in new fields. One of the potential advantages of micro-sized service-oriented technology companies, such as the one studied here, is that they are situated between suppliers and their customers, thereby being able to engage both upstream (suppliers) and downstream (their customers). The focal firm can influence the decision-making of customers by giving specific advice, for instance, on the lifespan of hardware and the potential for repair, that go beyond just short-term profitability" (Heyes et al., 2018)
Jensen et al.	2019	Journal of Cleaner Production	Three business case studies on remanufacturing at Philips Healthcare Refurbished Systems (RS), Siemens Wind Power (SWP), and Orangebox. The scope of the study is limited to OEMs	"The cases show that remanufacturing has the potential to create sustainable value across triple-bottom-line indicators (i.e., economic, environmental, social) by increasing sales, competitiveness, brand value, offering new sales opportunities, reducing environmental impacts of offered products through efficiency increase and material reduction, and creating new jobs

(continued)

Table 3 (continued)

Authors	Year of publication	Journal	Sample	Key findings
				and employment possibilities" (Jensen et al., 2019)
Leising et al.	2018	Journal of Cleaner Production	Three cases in the Dutch building sector	"Stakeholder collaboration is seen as essential for sustainable business model innovation, as well as for developing circular value chains. Moreover, incentives can guard and ground collaboration in the business model by deliberately incorporating (financial) interests of all stakeholders towards the collective goal of a circular building. Financial rewards are connected to the achievement of the collective goal and could help secure supply chain collaboration over even longer time frames" (Leising et al., 2018)
Manninen et al.	2018	Journal of Cleaner Production	Three business model cases – Case 1 represents a CE product-based BM in Finland, while cases 2 and 3 represent two types of CE service models, established in Finland and the Netherlands	"Assessment methods for CEBMs have to be chosen carefully and it should be ensured that results truly reflect defined value propositions. For instance, LCA (according to ISO 14040:2006), being developed for products and services, not for business models, is not always suitable and it is often too time- or resource-demanding for companies, but a streamlined LCA may be feasible in some cases. This means that other approaches are needed. For example, the assessment can be conducted by defining suitable quantitative or qualitative indicators, which set the criteria for value

				realization. In some cases, qualitative methods may be an alternative where a numeric calculation is impractical" (Manninen et al., 2018)
Ranta et al.	2018	Journal of Cleaner Production	Four cases from different geographic locations and employing different types of CE initiatives were chosen. A purposive sampling strategy was applied	"Take-back services enable the acquisition of particular wastes as resources, but they need to be incentivized through reductions in customers' total waste management costs. In each case, cost efficiency is driven by a take-back system ensuring that the focal firm can acquire waste suitable for efficient recycling. The take-back system for gaining economic value through CE can be implemented successfully in multiple ways. Implementation of the take-back system internally, through partnerships, or through purchasing from markets all appear to be applicable and successful approaches. At the same time recycling is easier to implement than reducing or reusing due to its smaller impact on the business model. Compared to reuse, where new activities such as refurbishment, maintenance, and remanufacturing, as well as separate sales of new and used products, are required for implementation; introducing recycling merely replaces virgin materials with recycled materials, with little effect on the fundamentals of the business model (e.g., target customers or revenue sources)" (Ranta et al., 2018)

(continued)

Table 3 (continued)

Authors	Year of publication	Journal	Sample	Key findings
Schwanholz and Leipold	2020	Journal of Cleaner Production	73 digital sharing platforms	"There are two factors crucial for the emergence of certain digital sharing practices. First, the presence of certain actors, e.g. large corporations, seems to influence the business models and motivations of digital sharing platforms. Second, the business models that digital sharing platforms take on and the motivations they pursue is likely related to their analogous 'heritage'" (Schwanholz & Leipold, 2020)
Urbinati et al.	2017	Journal of Cleaner Production	A total of 86 firms operating in different sectors of activity, which have adopted to different extents the principles of CE	"A new taxonomy of CEBMs that explains different degrees of circularity in practice adopted by firms in different industries based on two major dimensions: (a) the customer value proposition & interface, i.e., the implementation of the circularity concept in proposing value to customers; (b) the value network, i.e., the ways through which they interact with suppliers and reorganize their own internal activities" (Urbinati et al., 2017)
Urbinati et al.	2020	Journal of Cleaner Production	Three case studies. Cases A and B are two small-medium sized enterprises, whereas Case C is a technological center	"External environmental conditions, especially that of environmental regulation, can influence how companies implement the managerial practices for a circular business model and, in a different way, for the specific business model dimensions and related industrial benefits" (Urbinati et al., 2020)

Veleva and Bodkin	2018	Journal of Cleaner Production	Twelve US-based companies—Six entrepreneurs focused on product reuse or waste repurposing and six large companies with commitments to zero waste and/or circular business practices	"Companies launching CEBMs must offer easy and convenient ways to handle waste and surplus products to overcome inertia from business- as-usual practices and internal resistance to change. In a CE, developing a strong marketing strategy and close relationship with suppliers and customers is of critical importance as entrepreneurs have to ensure both a constant supply of waste/old products and constant demand for CE products/services. While the Internet and social media have made it easier than ever to buy and sell online, establishing close relationships requires trust, quality products and services, and competitive pricing. The study demonstrates that policymakers have a key role to play in advancing CE practices by a) enacting effective regulations or eliminating regulatory hurdles to CE practices; b) providing incentives to companies engaged in such practices; c) providing financial support; and d) raising awareness about the issue" (Veleva & Bodkin, 2018)
Werning and Spinler	2020	Journal of Cleaner Production	Semi-structured workshops and interviews, observations and discussions on the shop floor, and external and internal available data provided by Ricoh	"Helping existing businesses to handle the transition to CE can be done through effective and efficient management of their resources. In the process a large number of individual barriers may potentially have to be handled and prioritization is essential. Deploying resources to overcome the most threatening barriers for the business is a valid first step" (Werning & Spinler, 2020)

(Python built-in libraries), contextualized topic models (ZeroShotTM and RapidsAI in Python), and data processing of keywords found during ngram calculation using text preprocessing and vectorization (Andersen, 2020). Text preprocessing and vectorization was used to rank the strength of association between ngrams and the articles within which they appear, using an NLP neural network. Results were written using articles with the strongest association to the ngrams, while ngram frequencies were visualized with word clouds (Python WordCloud library). Contextualized topic models are presented as clusters of five words (see Table 4) that most often appear near each other throughout the corpus of articles. As opposed to the ngrams, clusters were trained using artificial neural nets. In addition, keywords in clusters do not necessarily occur right next to each other, as is the case with the ngrams.

The word clouds for ngrams were divided into three groups; the first group was most frequent ngrams (keywords that appear 500–2000 times within the corpus of documents), then less frequent ngrams (those that appear 100–500 times within the corpus of documents), and finally, ngrams of authors that appeared most frequently (authors that appeared 100–2000 times within the corpus of documents).

According to Türkeli et al. (2018), the most active research centers regarding CE are China (Beijing, Shenyang, Dalian, and Shanghai) and, in Europe, the Delft University of Technology in the Netherlands. In turn, the *Journal of Cleaner Production* is the most important journal for publications in regard to CE in both China and Europe and CE publications in general. The Ellen MacArthur Foundation also appeared frequently in the corpus of texts. Schulte (2013, p. 44) mentioned that the Ellen MacArthur Foundation published the first studies "highlighting the economic rationale for a transition to a CE." The quantitative analysis of bigrams and trigrams showed that raw materials play an essential role in CEBMs, particularly regarding the need for renewable raw materials. Companies with CEBMs are increasingly focusing on raw materials such as biodegradable packaging, recyclable wood composites, and recyclable textiles, among other things. This ties into the value proposition, waste management, model innovation, clean production, and recyclability ngrams (see Fig. 6), for example, competitive advantage for partners, additional incomes for industrial partners, reducing environmental stress in production, reducing expenses, creating new jobs, preserving nature and reducing wastes, etc. (D'Amato et al., 2020). The product and the design ngrams are focused on the bigram product design, which according to los Rios and Charnley (2017), in the context of CE, is related to efforts to create a closed-loop supply chain and shared ownership models. Products are thus designed for multiple life cycles (another keyword). The Circularity Indicators Project is a way to measure how well a product or company performs within the context of CE (los Rios & Charnley, 2017).

Since there are too many ngrams to be covered in this book chapter (see Fig. 7), only those mentioned in the chapter thus far will be further described. One of the benefits of data mining literature processing on CEBMs is providing additional support to the conclusions present in the qualitative literature review. One of the major focuses in a closed-loop chain versus a linear chain is that of reverse logistic. Rogers and Tibben-Lembke (1999) describe the process of reverse logistics as

Table 4 Literature review on circular economy and business models

Cluster	Keywords	Description
1	System, systems, service, PSS, services	Product-Service Systems
2	Material, materials, raw, resources, loops	Raw material resource loops
3	Sustainability, development, sustainable, green, growth	Sustainable development (green) and growth
4	Industry, sector, industrial, concept, fashion	Fashion industry
5	Product, services, products, service, consumer	Customer product services
6	Plastic, plastics, waste, recycling, material	Waste and material recycling
7	Market, new, bm, flexzhouse, housing	Flexzhouse housing market
8	Consumption, emissions, natural, energy, carbon	Energy consumption and carbon emission
9	Remanufacturing, activities, equipment, logistics, maintenance	Activities, equipment, and logistics in remanufacturing and maintenance
10	Environmental, social, environment, economic, impact	Environmental and social impact
11	Value, creation, network, proposition, capture	Value proposition creation, capture, and network
12	Supply, life, cycle, chain, loop	Life cycle, supply chain loop
13	Design, framework, project, strategies, approach	Project management
14	Innovation, model, models, sustainable, business	Innovative sustainable business models
15	Food, energy, supply, processes, consumption	Food supply and energy processes and consumption
16	Waste, management, recycling, solid, conservation	Waste management
17	al, et, bocken, van, foundation,	Bocken et al.
18	Data, information, analysis, collection, big	Big data analysis
19	CE, strategies, implementation, strategy, principles	CE implementation strategies and principles
20	Foundation, towards, Ellen, Macarthur, economy	Ellen Macarthur Foundation

"planning, implementing, and controlling the efficient, cost-effective flow of raw materials, in-process inventory, finished goods, and related information from the point of consumption to the point of origin to recapture value or proper disposal" (de Brito & Dekker, 2004). Industrial ecology (IE) was also mentioned in the beginning of this chapter and was also one of the keywords. IE can be divided into four categories: industrial symbiosis, industrial ecosystem, industrial metabolism, and environmental legislation and regulations. IE is an interdisciplinary field of study that looks at different methods, approaches, and frameworks for developing

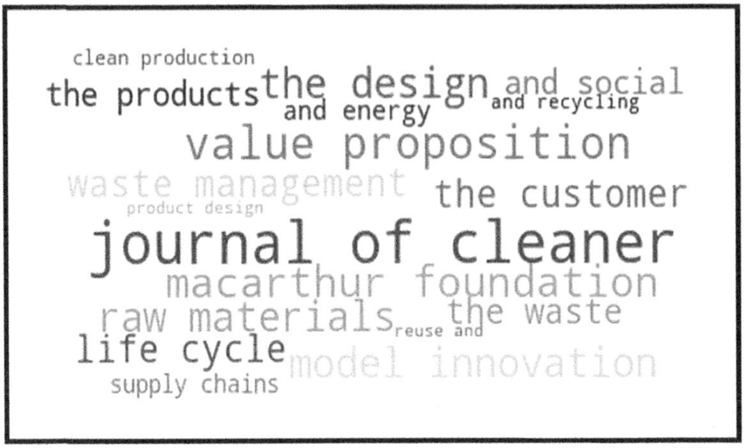

Fig. 6 Most frequent ngrams (appear 500–2000 times)

closed-loop industrial ecosystems (Li, 2018). IE looks at industrial systems and compares them with biological ecosystems in order to study material energy metabolism (industrial metabolism) and the symbiotic relationships among different organisms (industrial symbiosis) and to figure out how these findings can be applied to our current industrial systems (Han et al., 2021). However, industrial symbiosis (IS) and industrial metabolism (IM) differ from common CE strategies. IS and IM look at the relationships among companies in how they use resources and handle waste, while CEBMs tend to look at PSS or the shared economy and focus on the customer instead of the interconnectedness between companies (Fraccascia et al., 2019).

Product-service systems (PSS) is another ngram that appears very frequently in this literature review and quantitative analysis of articles. As was mentioned earlier, the most recurrent PSS types adopted to enable circular strategies are the product renting/sharing, lease, and product-related service. According to Mair and Reischauer (2017), the sharing economy, which is another keyword, uses digital platforms provided by a company, allowing individuals to access goods and services shared with other individuals, thereby decreasing the need for each individual to buy and own various resources. Product-service systems (PSS) can be realized by the use of smart products that like one presented by Pagoropoulos et al. (2017) can use technology solutions to track material flows for value recovery, using the so-called Re-strategies (Reuse, Repair, and Remanufacture). This, in turn, helps achieve the IE goal of transitioning to closed-loop systems. These closed-loop systems are connected through the Internet of things, where sensors collect big data and use it to convey specific life cycle information to the stakeholder (data-to-service). Big data analysis can be used in PSS to enable the monitoring of product life cycles; this is called a Product Life cycle Management (PLM) system. An example of data passed through PLM systems is information on which components and materials can

Fig. 7 Less frequent ngrams (appear 100–500 times)

be recycled or re-used at the end of the product's shelf life, enabling stock and material flow near time consumption and optimization. Other examples of big data analytics that can increase manufacturing sustainability are the maintenance of internal combustion engines, waste management, adaptive calibration, monitoring production and consumption processes, the re-use of spare parts, and engine remanufacturing, among other things.

5 Discussion with Conclusions

The added value of this chapter is a systematic literature review and, in addition, bibliometric and topic analysis, which complements and strengthens findings from the systematic literature review. For the systematic literature review, we have taken four steps, which led us to a thoughtful review of 201 papers altogether, and due to several criteria, we identified 51 papers as the final pool. The final pool of 51 papers were independently scrutinized by two researchers and are presented in more detail in Table 3 (Sect. 3, "Systematic Literature Review: Methodology and Results"). Altogether, out of the 51 papers, 15 were published in the *Journal of Cleaner Production*, followed by 14 papers published in *Resources, Conservation & Recycling*, and 12 papers in *Procedia CIRP*. In addition, two papers were published in *Ecological Economics*, two in *CIRP Journal of Manufacturing Science and Technology*, one in *Resources Policy*, one in *Waste Management*, one in *Forest Policy and Economics*, one in *Geoforum*, one in *Journal of Business Research*, and one in *Business Horizons*.

After reviewing these papers, we came to a threefold conclusion. First, most papers are theoretical and therefore not presented in the abovementioned table; however, they are involved in the introduction of this chapter. We can also see that the first publications of empirical research linking the CE and business models start in 2017 with a clear increase in 2018 and especially in 2020. Second, empirical papers are qualitative, which means that research is conducted as case studies, interviews, and secondary data or focus groups, and not primarily conducted as quantitative research with a survey on a large scale (only three papers identified). Third, the topics that the reviewed papers cover and mention as their findings are, in the majority, servitization, product-service systems (PSS), Circular Manufacturing Systems (CMS) approach, service-based business models (SBBM), IoT, and the importance of sophisticated ICT. For example, Pieroni et al. (2018) explored how different PSS types can enable several circular strategies for resource productivity. The main trends are as follows: PSS appears to be conducive to the synergistic application of multiple circular strategies; combining resource efficiency strategies (preventative) with effective resource strategies (for extending products/materials life) seems to be a trend in companies implementing PSS; circular solutions based on the result-oriented PSS are less frequent, although the potentials for energy savings and resource decoupling are claimed to be higher; and companies have room to improve the application of more diversified CE strategies such as new technology paradigms (refuse), rethinking value propositions and offerings' fulfillment (rethink), and extending the life span of products through refurbishment and remanufacturing.

Research has also found that digitalization can boost the transformation toward a more sustainable CE in a way that it can help close the material loops by providing accurate information on the availability, location, and condition of products (Pagoropoulos et al., 2017). Product life (keyword), or durability, in other words, is said to be among the most important ways to reduce waste (lower resource consumption). The product life span can be influenced by factors that fall under three categories: product functionality, consumer expectations, and product obsolescence. These three categories are primarily focused on the durability of the product, reversing obsolescence, and the consumers' desired durability of products (Singh et al., 2019).

Since the first three clusters, PSS, raw materials, and sustainable, have already been talked about (see Table 4), we will go forward to cluster 4 (Fashion industry). In the fashion industry, there are macro trends shaping sustainable fashion business models (Todeschini et al., 2017); these are consumer awareness (capsule wardrobe, lowsumerism, slow fashion), CE (recycling, vegan, upcycling), corporate social responsibility (sweatshop-free, fair trade, locally sourced), sharing economy and collaborative consumption (fashion library, second hand, collaboration), and technological innovation (sustainable raw materials, zero waste, wearables). Since customer product services, waste, and material recycling (clusters 5 and 6) have already been discussed, we will continue to cluster 7, which is the Flexzhouse housing market in Malaysia. This was detected as a less frequent ngram (see Table 4) because Noor (2017) wrote a 318-page thesis on sustainable housing in Malaysia. This is over a dozen times as big as a typical scientific article, meaning that it compares to several articles talking about the topic. The remaining clusters that have not already been covered are cluster 8, cluster 15, cluster 17, cluster 19, and cluster 20. Cluster 8 is mainly concerned with carbon emissions. The major thing to point out with energy consumption and carbon emissions is that this is considered when talking about greenhouse gases (GHG), which are responsible for 55% of all emissions in the world. These emissions need to be lowered by 11.3% annually by the year 2050; currently, the number is seven times higher than that (Durán-Romero et al., 2020). Cluster 15 deals with food supply, energy processes, and consumption and is related to moving food production closer to consumption, better shelf management, and solutions that optimize protection using less packaging material. With food processing, the idea is to return waste as nutrients back into our natural ecosystem; for example, Germany is already separating meat from other food waste and then using it as compost (Schulte, 2013). For cluster 19, the principles and implementation strategies can be summarized by the Ellen MacArthur Foundation (2015) CE framework called ReSOLVE (Jabbour et al., 2020a, b). This consists of the following principles: Regenerate (renewable and biodegradable sources), Share (durability and product as a service), Optimize (energy efficiency, decrease of waste, PSS), Loop (repairing, re-using, refurbishing, remanufacturing, and recycling of resources), Virtualize (digital services), and Exchange (disruptive technologies).

Fig. 8 Ngrams of authors that frequently appear in text

However, these principles have several barriers, such as Structural (communication difficulties), Operational (lack of operational capacity and infrastructure), Attitudinal (risk aversion), and Technological (difficulty acquiring or implementing technologies). Cluster 20 relates to the Ellen MacArthur Foundation. The foundation has been mentioned a few times in previous paragraphs; however, not much has been said about what the foundation does. According to the Ellen MacArthur Foundation (2021), the foundation creates evidence-based research on the topic of CE and educates teachers and academics through formal learning opportunities. They also work with "businesses, international institutions, governments, cities, universities, non-governmental organizations, innovators, and many others," as well as creating networks for members of the foundation to communicate and learn from each other. The Ellen MacArthur Foundation (2021) focuses mostly on areas that they believe will have the biggest impact when shifting to the CE: plastics, food, fashion, finance, and cities. We left cluster 17 for last, given that it is related to a particular group of ngrams that were found during the analysis. The following word cloud displays several authors that were cited several times in the articles that were scraped. "Bocken et al." appears in the ngrams 972 times, often also as an author of the articles most associated with a particular ngram. "Bocken et al." is in reference to Nancy Bocken from Maastricht University. In addition to Nancy Bocken, several other authors appear as ngrams between 100 and 400 times (see Fig. 8).

Webscraping and data processing done on 407 open-source articles on the topic of circular business models mostly confirmed our findings from the literature review. Among the concepts most often correlated to the topic of CEBM are the *Journal of Cleaner Production*, the Ellen MacArthur Foundation, value proposition, model innovation, waste and waste management, supply chain, clean production, life cycle, products, and product design. Companies with CEBMs are increasingly focusing on design as a first step toward products and clean production processes that bring competitive advantages for partners, not only consumers but also other stakeholders, including the environment. This is linked to the value proposition, waste management, BM innovation, and product recyclability, all focused on creating a closed-loop supply chain where the products and resources with which they are made are designed for multiple life cycles. In addition, production involves the use of sustainable energy sources that limits the use of non-renewable resources. All this is supported with the use of new digital technologies and data platforms to fully decouple value creation from the consumption of finite resources. In fact, these new solutions are supported by the policies of the European Commission that seek economic growth together with the minimization of environmental impact.

References

Aid, G., Eklund, M., Anderberg, S., & Baas, L. (2017). Expanding roles for the Swedish waste management sector in inter-organizational resource management. *Resources, Conservation and Recycling, 124*, 85–97. https://doi.org/10.1016/j.resconrec.2017.04.007

Antikainen, M., Uusitalo, T., & Kivikytö-Reponen, P. (2018). Digitalisation as an enabler of circular economy. *Procedia CIRP, 7*, 45–49. https://doi.org/10.1016/j.procir.2018.04.027

Andersen, S. (2020). Natural language processing: text preprocessing and vectorizing at rocking speed with RAPIDS cuML https://medium.com/rapids-ai/natural-language-processing-text-preprocessing-and-vectorizing-at-rocking-speed-with-rapids-cuml-74b8d751812e. Accessed 21 Nov. 2020.

Asif, F. M. A., Roci, M., Lieder, M., Rashid, A., Štimulak, M., Halvordsson, E., & De Bruijckere, R. (2018). A practical ICT framework for transition to circular manufacturing systems. *Procedia CIRP, 72*, 598–602. https://doi.org/10.1016/j.procir.2018.03.311

Åkerman, M., Humalisto, N., & Pitzen, S. (2020). Material politics in the circular economy: The complicated journey from manure surplus to resource. *Geoforum, 116*, 73–80. https://doi.org/10.1016/j.geoforum.2020.07.013

Bauwens, T., Hekkert, M., & Kirchherr, J. (2020). Circular futures: What will they look like? *Ecological Economics, 175*, 106703. https://doi.org/10.1016/j.ecolecon.2020.106703

Bocken, N. M., Short, S. W., Rana, P., & Evans, S. (2014). A literature and practice review to develop sustainable business model archetypes. *Journal of Cleaner Production, 65*, 42–56. https://doi.org/10.1016/j.jclepro.2013.11.039

Bressanelli, G., Perona, M., & Saccani, N. (2017). Reshaping the washing machine industry through circular economy and product-service system business models. *Procedia CIRP, 64*, 43–48. https://doi.org/10.1016/j.procir.2017.03.065

Bressanelli, G., Adrodegari, F., Perona, M., & Saccani, N. (2018). The role of digital technologies to overcome circular economy challenges in PSS business models: An exploratory case study. *Procedia CIRP, 73*, 216–221. https://doi.org/10.1016/j.procir.2018.03.322

Copani, G., & Behnam, S. (2020). Remanufacturing with upgrade PSS for new sustainable business models. *CIRP Journal of Manufacturing Science and Technology, 29*, 245–256. https://doi.org/10.1016/j.cirpj.2018.10.005

D'Amato, D., Veijonaho, S., & Toppinen, A. (2020). Towards sustainability? Forest-based circular bioeconomy business models in Finnish SMEs. *Forest Policy Economy, 110*, 101848. https://doi.org/10.1016/j.forpol.2018.12.004

De Angelis, R., & Feola, R. (2020). Circular business models in biological cycles: The case of an Italian spin-off. *Journal of Cleaner Production, 247*, 119603. https://doi.org/10.1016/j.jclepro.2019.119603

de Brito, M., & Dekker, R. (2004). Chapter 1. A framework for reverse logistics. In *Quantitative models for closed-loop supply chains* (pp. 3–27). Springer.

De Pádua Pieroni, M., Blomsma, F., McAloone, T. C., & Pigosso, D. C. A. (2018). Enabling circular strategies with different types of product/service-systems. *Procedia CIRP, 73*, 179–184. https://doi.org/10.1016/j.procir.2018.03.327

Diaz Lopez, F. J., Bastein, T., & Tukker, A. (2019). Business model innovation for resource-efficiency, circularity and cleaner production: What 143 cases tell us. *Ecological Economics, 155*, 20–35. https://doi.org/10.1016/j.ecolecon.2018.03.009

Durán-Romero, G., López, A. M., Beliaeva, T., Ferasso, M., Garonne, C., & Jones, P. (2020). Bridging the gap between circular economy and climate change mitigation policies through eco-innovations and quintuple helix model. *Technological Forecasting and Social Change, 160*, 120246. https://doi.org/10.1016/j.techfore.2020.120246

Eguren, J. A., Justel, D., Iriarte, I., & Esnaola, A. (2018). Opportunities and incentives for remanufacturing in the Basque Country. *Procedia CIRP, 73*, 253–258. https://doi.org/10.1016/j.procir.2018.04.081

Ellen MacArthur Foundation. (2015). Towards the circular economy. Economic and business rationale for an accelerated transition. Accessed online: https://www.werktrends.nl/app/uploads/2015/06/Rapport_McKinsey-Towards_A_Circular_Economy.pdf (4. 8. 2021).

Ellen MacArthur Foundation. (2021). About us—what we do: How to build a circular economy. Accessed online: https://ellenmacarthurfoundation.org/about-us/what-we-do (16.11.2021).

Ertz, M., Leblanc-Proulx, S., Sarigöllü, E., & Morin, V. (2019). Advancing quantitative rigor in the circular economy literature: New methodology for product lifetime extension business models. *Resources, Conservation and Recycling, 150*, 104437. https://doi.org/10.1016/j.resconrec.2019.104437

Fraccascia, L., Giannoccaro, I., & Albino, V. (2019). Business models for industrial symbiosis: A taxonomy focused on the form of governance. *Resources, Conservation and Recycling, 146*, 114–126. https://doi.org/10.1016/j.resconrec.2019.03.016

Esmaeilian, B., Wang, B., Lewis, K., Duarte, F., Ratti, C., & Behdad, S. (2018). The future of waste management in smart and sustainable cities: A review and concept paper. *Waste Management, 81*, 177–195. https://doi.org/10.1016/j.wasman.2018.09.047

Gall, M., Wiener, M., Chagas de Oliveira, C., Lang, R. W., & Hansen, E. G. (2020). Building a circular plastics economy with informal waste pickers: Recyclate quality, business model, and societal impacts. *Resources, Conservation and Recycling, 156*, 104685. https://doi.org/10.1016/j.resconrec.2020.104685

Geissdoerfer, M., Morioka, S. N., de Carvalho, M. M., & Evans, S. (2018). Business models and supply chains for the circular economy. *Journal of Cleaner Production, 190*, 712–721. https://doi.org/10.1016/j.jclepro.2018.04.159

Geissdoerfer, M., Pieroni, M. P. P., Pigosso, D. C. A., & Soufani, K. (2020). Circular business models: A review. *Journal of Cleaner Production, 277*, 123741. https://doi.org/10.1016/j.jclepro.2020.123741

Giampietro, M., & Funtowicz, S. O. (2020). From elite folk science to the policy legend of the circular economy. *Environmental Science and Policy, 109*, 64–72. https://doi.org/10.1016/j.envsci.2020.04.012

Gnoni, M. G., Mossa, G., Mummolo, G., Tornese, F., & Verriello, R. (2017). Supporting circular economy through use-based business models: The washing machines case. *Procedia CIRP, 64*, 49–54. https://doi.org/10.1016/j.procir.2017.03.018

Guldmann, E., & Huulgaard, R. D. (2020). Barriers to circular business model innovation: A multiple-case study. *Journal of Cleaner Production, 243*, 118160. https://doi.org/10.1016/j.jclepro.2019.118160

Gusmerotti, N. M., Testa, F., Corsini, F., Pretner, G., & Iraldo, F. (2019). Drivers and approaches to the circular economy in manufacturing firms. *Journal of Cleaner Production, 230*, 314–327. https://doi.org/10.1016/j.jclepro.2019.05.044

Guzzo, D., Carvalho, M. M., Balkenende, R., & Mascarenhas, J. (2020). Circular business models in the medical device industry: Paths towards sustainable healthcare. *Resources, Conservation and Recycling, 160*, 104904. https://doi.org/10.1016/j.resconrec.2020.104904

Haanstra, W., Toxopeus, M. E., & Van Gerrevink, M. R. (2017). Product life cycle planning for sustainable manufacturing: Translating theory into business opportunities. *Procedia CIRP, 61*, 46–51. https://doi.org/10.1016/j.procir.2016.12.005

Han, F., Feng, Z., Wang, C., Yang, N., & Yang, D. (2021). Interweaving industrial ecology and ecological modernization: A comparative bibliometric analysis. *Sustainability, 13*, 9673. https://doi.org/10.3390/su13179673

He, T., Ho, W., Zhang, Y., & Dey, P. K. (2016). Organising the business processes of a product servitised supply chain: A value perspective. *Production Planning & Control, 27*, 378–393. https://doi.org/10.1080/09537287.2015.1128571

Henry, M., Bauwens, T., Hekkert, M., & Kirchherr, J. (2020). A typology of circular start-ups: Analysis of 128 circular business models. *Journal of Cleaner Production, 245*, 118528. https://doi.org/10.1016/j.jclepro.2019.118528

Heyes, G., Sharmina, M., Mendoza, J. M. F., Gallego-Schmid, A., & Azapagic, A. (2018). Developing and implementing circular economy business models in service-oriented technology companies. *Journal of Cleaner Production, 177*, 621–632. https://doi.org/10.1016/j.jclepro.2017.12.168

Hopkinson, P., De Angelis, R., & Zils, M. (2020). Systemic building blocks for creating and capturing value from circular economy. *Resources, Conservation and Recycling, 155*, 104672. https://doi.org/10.1016/j.resconrec.2019.104672

Ingemarsdotter, E., Jamsin, E., & Balkenende, R. (2020). Opportunities and challenges in IoT-enabled circular business model implementation—A case study. *Resources, Conservation and Recycling, 162*, 105047. https://doi.org/10.1016/j.resconrec.2020.105047

Jabbour, C. J., De Camargo Fiorini, P., Wong, C. W. Y., Jugend, D., Lopes De Sousa Jabbour, A. B., Roman Pais Seles, B. M., Paula Pinheiro, M. A., & Ribeiro da Silva, H. M. (2020a). First-mover firms in the transition towards the sharing economy in metallic natural resource-intensive industries: Implications for the circular economy and emerging industry 4.0 technologies. *Resources Policy, 66*, 101596. https://doi.org/10.1016/j.resourpol.2020.101596

Jabbour, C. J. C., Seuring, S., Jabbour, A. B. L. S., Jugend, D., Fiorini, P. D. C., Latan, H., & Izeppi, W. C. (2020b). Stakeholders, innovative business models for the circular economy and sustainable performance of firms in an emerging economy facing institutional voids. *Journal of Environmental Management, 264*, 1104–1116. https://doi.org/10.1016/j.jenvman.2020.110416

Jensen, J. P., Prendeville, S. M., Bocken, N. M. P., & Peck, D. (2019). Creating sustainable value through remanufacturing: Three industry cases. *Journal of Cleaner Production, 218*, 304–314. https://doi.org/10.1016/j.jclepro.2019.01.301

Jesson, J. K., Matheson, L., & Lacey, F. M. (2011). *Doing your literature review. Traditional and systematic techniques*. SAGE Publications.

Khan, M. A., West, S., & Wuest, T. (2020). Midlife upgrade of capital equipment: A servitization-enabled, value-adding alternative to traditional equipment replacement strategies. *CIRP Journal of Manufacturing Science and Technology, 29*, 232–244. https://doi.org/10.1016/j.cirpj.2019.09.001

Kirchherr, J., Reike, D., & Hekkert, M. (2017). Conceptualizing the circular economy: An analysis of 114 definitions. *Resources, Conservation & Recycling, 127*, 221–232. https://doi.org/10.1016/j.resconrec.2017.09.005

Klenk, F., Gleich, K., Meister, F., Häfner, B., & Lanza, G. (2020). Approach for developing implementation strategies for circular economy in global production networks. *Procedia CIRP, 90*, 127–132. https://doi.org/10.1016/j.procir.2020.01.052

Kobayashi, H., Murata, H., & Fukushige, S. (2020). Connected life-cycle systems: A new perspective on industrial symbiosis. *Procedia CIRP, 90*, 388–392. https://doi.org/10.1016/j.procir.2020.01.107

Konietzko, J., Baldassarre, B., Brown, P., Bocken, N., & Hultink, E. J. (2020). Circular business model experimentation: Demystifying assumptions. *Journal of Cleaner Production, 277*, 122596. https://doi.org/10.1016/j.jclepro.2020.122596

Kravchenko, M., McAloone, T. C., & Pigosso, D. C. A. (2020). To what extent do circular economy indicators capture sustainability? *Procedia CIRP, 90*, 31–36. https://doi.org/10.1016/j.procir.2020.02.118

Krystofik, M., Luccitti, A., Parnell, K., & Thurston, M. (2018). Adaptive remanufacturing for multiple life-cycles: A case study in office furniture. *Resources, Conservation and Recycling, 135*, 14–23. https://doi.org/10.1016/j.resconrec.2017.07.028

Leder, N., Kumar, M., & Rodrigues, V. S. (2020). Influential factors for value creation within the circular economy: Framework for waste valorisation. *Resources, Conservation and Recycling, 158*. https://doi.org/10.1016/j.resconrec.2020.104804

Leising, E., Quist, J., & Bocken, N. (2018). Circular economy in the building sector: Three cases and a collaboration tool. *Journal of Cleaner Production, 176*, 976–989. https://doi.org/10.1016/j.jclepro.2017.12.010

Li, X. (2018). Industrial ecology and industrial Symbiosis - definitions and development histories. In *Industrial ecology and industry symbiosis for environmental sustainability*. Palgrave Pivot. https://doi.org/10.1007/978-3-319-67501-5_2

Lieder, M., Asif, F. M. A., & Rashid, A. (2017). Towards circular economy implementation: An agent-based simulation approach for business model changes. *Autonomous Agents and Multi-Agent Systems, 31*(6), 1377–1402. https://doi.org/10.1007/s10458-017-9365-9

los Rios, I. C. D., & Charnley, F. J. S. (2017). Skills and capabilities for a sustainable and circular economy: The changing role of design. *Journal of Cleaner Production, 160*, 109–122.

Mair, J., & Reischauer, G. (2017). Capturing the dynamics of the sharing economy: Institutional research on the plural forms and practices of sharing economy organizations. *Technological Forecasting and Social Change*. https://doi.org/10.1016/j.techfore.2017.05.023

Maffei, A., Grahn, S., & Nuur, C. (2019). Characterization of the impact of digitalization on the adoption of sustainable business models in manufacturing. *Procedia CIRP, 81*, 765–770. https://doi.org/10.1016/j.procir.2019.03.191

Manninen, K., Koskela, S., Antikainen, R., Bocken, N., Dahlbo, H., & Aminoff, A. (2018). Do circular economy business models capture intended environmental value propositions? *Journal of Cleaner Production, 171*, 413–422. https://doi.org/10.1016/j.jclepro.2017.10.003

Moreno, M., Turner, C., Tiwari, A., Hutabarat, W., Charnley, F., Widjaja, D., & Mondini, L. (2017). Re-distributed manufacturing to achieve a circular economy: A case study utilizing IDEF0 Modeling. *Procedia CIRP, 63*, 686–691. https://doi.org/10.1016/j.procir.2017.03.322

Muranko, Z., Andrews, D., Chaer, I., Newton, E. J., Proudman, P., & Longhurst, M. (2017). Pro-circular behaviours and refrigerated display cabinets: Supporting resource efficiency in the retail refrigeration sector. *Energy Procedia, 123*, 70–75. https://doi.org/10.1016/j.egypro.2017.07.235

Muranko, Z., Andrews, D., Chaer, I., & Newton, E. J. (2019). Circular economy and behaviour change: Using persuasive communication to encourage pro-circular behaviours towards the purchase of remanufactured refrigeration equipment. *Journal of Cleaner Production, 222*, 499–510. https://doi.org/10.1016/j.jclepro.2019.02.219

Noor, M. Z. (2017). FlexZhouse: New business model for affordable housing in Malaysia. 2. 1–318.

Nußholz, J. L. K., Nygaard Rasmussen, F., & Milios, L. (2019). Circular building materials: Carbon saving potential and the role of business model innovation and public policy. *Resources, Conservation and Recycling, 141*, 308–316. https://doi.org/10.1016/j.resconrec.2018.10.036

Osterwalder, A., Pigneur, Y., Clark, T., & Smith, A. (2010). *Business model generation: A handbook for visionaries, game changers, and challengers*. Wiley.

Pagoropoulos, A., Pigosso, D. C. A., & McAloone, T. C. (2017). The emergent role of digital technologies in the circular economy: A review. *Procedia CIRP, 64*, 19–24. https://doi.org/10.1016/j.procir.2017.02.047

Panarotto, M., Wall, J., & Larsson, T. (2017). Simulation-driven design for assessing strategic decisions in the conceptual design of circular PSS business models. *Procedia CIRP, 64*, 25–30. https://doi.org/10.1016/j.procir.2017.03.026

Parida, V., Burström, T., Visnjic, I., & Wincent, J. (2019). Orchestrating industrial ecosystem in circular economy: A two-stage transformation model for large manufacturing companies. *Journal of Business Research, 101*, 715–725. https://doi.org/10.1016/j.jbusres.2019.01.006

Pieroni, M., Blomsma, F., McAloone, T. C., & Pigosso, D. C. A. (2018). Enabling circular strategies with different types of product/service-systems. *Procedia CIRP, 73*, 179–184. https://doi.org/10.1016/j.procir.2018.03.327

Pieroni, M. P. P., McAloone, T. C., & Pigosso, D. C. A. (2019). Business model innovation for circular economy and sustainability: A review of approaches. *Journal of Cleaner Production, 215*, 198–216. https://doi.org/10.1016/j.jclepro.2019.01.036

Pieroni, M. P. P., McAloone, T. C., & Pigosso, D. C. A. (2020). From theory to practice: Systematising and testing business model archetypes for circular economy. *Resources, Conservation and Recycling, 162*, 105029. https://doi.org/10.1016/j.resconrec.2020.105029

Preston, F. (2012). A global redesign? Shaping the circular economy. *Energy, Environment and Resource Governance., 2*, 1–20.

Primc, K., Kalar, B., Slabe-Erker, R., Dominko, M., & Ogorevc, M. (2020). Circular economy configuration indicators in organizational life cycle theory. *Ecological Indicators, 116*, 106532. https://doi.org/10.1016/j.ecolind.2020.106532

Ranta, V., Aarikka-Stenroos, L., & Mäkinen, S. J. (2018). Creating value in the circular economy: A structured multiple-case analysis of business models. *Journal of Cleaner Production, 201*, 988–1000. https://doi.org/10.1016/j.jclepro.2018.08.072

Rogers, D. S., & Tibben-Lembke, R. S. (1999). *Going backwards: Reverse logistics trends and practices*. Reverse Logistics Executive Council.

Saidani, M., Yannou, B., Leroy, Y., & Cluzel, F. (2018). Heavy vehicles on the road towards the circular economy: Analysis and comparison with the automotive industry. *Resources, Conservation and Recycling, 135*, 108–122. https://doi.org/10.1016/j.resconrec.2017.06.017

Sarc, R., Curtis, A., Kandlbauer, L., Khodier, K., Lorber, K. E., & Pomberger, R. (2019). Digitalisation and intelligent robotics in value chain of circular economy oriented waste management – A review. *Waste Management, 95*, 476–492. https://doi.org/10.1016/j.wasman.2019.06.035

Schulte, U. (2013). New business models for a radical change in resource efficiency. *Environmental Innovation and Societal Transitions, 9*, 43–47. https://doi.org/10.1016/j.eist.2013.09.006

Schwanholz, J., & Leipold, S. (2020). Sharing for a circular economy? An analysis of digital sharing platforms' principles and business models. *Journal of Cleaner Production, 269*, 1223–1227. https://doi.org/10.1016/j.jclepro.2020.122327

Singh, J., Cooper, T., Cole, C., Gnanapragasam, A., & Shapley, M. (2019). Evaluating approaches to resource management in consumer product sectors: An overview of global practices. *Journal of Cleaner Production, 224*, 218–237. https://doi.org/10.1016/j.jclepro.2019.03.203

Sousa-Zomer, T. T., Magalhães, L., Zancul, E., & Cauchick-Miguel, P. A. (2017). Lifecycle management of product-service systems: A preliminary investigation of a white goods manufacturer. *Procedia CIRP, 64*, 31–36. https://doi.org/10.1016/j.procir.2017.03.041

Sousa-Zomer, T. T., Magalhães, L., Zancul, E., & Cauchick-Miguel, P. A. (2018). Exploring the challenges for circular business implementation in manufacturing companies: An empirical investigation of a pay-per-use service provider. *Resources, Conservation and Recycling, 135*, 3–13. https://doi.org/10.1016/j.resconrec.2017.10.033

Stahel, W. R. (2020). History of the circular economy. The historic development of circularity and the circular economy. In S. Eisenriegler (Ed.), *The circular economy in the European Union (7–19)*. Springer.

Stewart, R., Niero, M., Murdock, K., & Olsen, S. I. (2018). Exploring the implementation of a circular economy strategy: The case of a closed-loop supply of aluminum beverage cans. *Procedia CIRP, 69*, 810–815. https://doi.org/10.1016/j.procir.2017.11.006

Todeschini, B. V., Cortimiglia, M. N., Callegaro-de-Menezes, D., & Ghezzi, A. (2017). Innovative and sustainable business models in the fashion industry: Entrepreneurial drivers, opportunities, and challenges. *Business Horizons, 60*, 759–770. https://doi.org/10.1016/j.bushor.2017.07.003

Tranfield, D., Denyer, D., & Smart, P. (2003). Towards a methodology for developing evidence-informed management knowledge by means of systematic review. *British Journal of Management, 14*(3), 207–222.

Türkeli, S., Kemp, R., Huang, B., Bleischwitz, R., & McDowall, W. (2018). Circular economy scientific knowledge in the European Union and China: A bibliometric, network and survey analysis (2006–2016). *Journal of Cleaner Production, 197*, 1244–1261. https://doi.org/10.1016/j.jclepro.2018.06.118

Uçar, E., Le Dain, M. A., & Joly, I. (2020). Digital technologies in circular economy transition: Evidence from case studies. *Procedia CIRP, 90*, 133–136. https://doi.org/10.1016/j.procir.2020.01.058

Ünal, E., Urbinati, A., Chiaroni, D., & Manzini, R. (2019). Value creation in circular business models: The case of a US small medium enterprise in the building sector. *Resources, Conservation and Recycling, 146*, 291–307. https://doi.org/10.1016/j.resconrec.2018.12.034

Urbinati, A., Chiaroni, D., & Chiesa, V. (2017). Towards a new taxonomy of circular economy business models. *Journal of Cleaner Production, 168*, 487–498. https://doi.org/10.1016/j.jclepro.2017.09.047

Urbinati, A., Rosa, P., Sassanelli, C., Chiaroni, D., & Terzi, S. (2020). Circular business models in the European manufacturing industry: A multiple case study analysis. *Journal of Cleaner Production, 274*, 122964. https://doi.org/10.1016/j.jclepro.2020.122964

Veleva, V., & Bodkin, G. (2018). Corporate-entrepreneur collaborations to advance a circular economy. *Journal of Cleaner Production, 188*, 20–37. https://doi.org/10.1016/j.jclepro.2018.03.196

Werning, J. P., & Spinler, S. (2020). Transition to circular economy on firm level: Barrier identification and prioritization along the value chain. *Journal of Cleaner Production, 245*, 118609. https://doi.org/10.1016/j.jclepro.2019.118609

Whalen, K. A., Milios, L., & Nussholz, J. (2018). Bridging the gap: Barriers and potential for scaling re-use practices in the Swedish ICT sector. *Resources, Conservation and Recycling, 135*, 123–131. https://doi.org/10.1016/j.resconrec.2017.07.029

Widmer, T., Tjahjono, B., & Bourlakis, M. (2018). Defining value creation in the context of circular PSS. *Procedia CIRP, 73*, 142–147. https://doi.org/10.1016/j.procir.2018.03.329

Witjes, S., & Lozano, R. (2016). Towards a more circular economy: Proposing a framework linking sustainable public procurement and sustainable business models. *Resources, Conservation & Recycling, 112*, 37–44. https://doi.org/10.1016/j.resconrec.2016.04.015

Zheng, P., Wang, Z., Chen, C. H., & Pheng Khoo, L. (2019). A survey of smart product-service systems: Key aspects, challenges and future perspectives. *Adv. Eng. Informatics, 42*, 100973. https://doi.org/10.1016/j.aei.2019.100973

Sustainable Circular Business Models: The Circular Economy Needs Responsibility and Purpose to Fulfil its Promises

Christoph H. Wecht, Beate Cesinger, Christine Vallaster, and Natalie Aleksić

Abstract *Circular economy*, as a concept of societal change towards a more balanced view of achieving the tripartite objectives of profit–people–planet, is a useful reference point for companies which are solely limited to a shareholder view. This goes hand in hand with the long overdue transformation from shareholder-capitalism to a stakeholder- and planetary-driven sustainable economy. However, circular economy per se does not guarantee the achievement of sustainability, as in most cases it is still economically driven. Only paired with responsibility and purpose (on the owner's side), means of circular economy result in sustainable circular business models, which we consider as the basis of a future economic system for the benefit of planet and society. Using five examples from the apparel and footwear industry, we develop a framework and three design propositions to illustrate archetypes and pathways how companies can transform their business models from purely economic-driven into sustainable circular business models (Remark: Parts of this chapter have been presented at The ISPIM Innovation Conference – Innovating Our Common Future, Berlin, Germany on 20–23 June 2021.).

C. H. Wecht (✉) · B. Cesinger
New Design University, St. Pölten, Austria
e-mail: christoph.wecht@ndu.ac.at; beate.cesinger@ndu.ac.at

C. Vallaster
Salzburg University of Applied Sciences, Puch/Salzburg, Austria
e-mail: christine.vallaster@fh-salzburg.ac

N. Aleksić
Stockholm School of Economics, Stockholm, Sweden
e-mail: natalie.aleksic@phdstudent.hhs.se

V. Prokop et al. (eds.), *Business Models for the Circular Economy*, Sustainability and Innovation, https://doi.org/10.1007/978-3-031-08313-6_3

1 Introduction

Circular economy can be perceived as an opportunity to meet current socio-ecological challenges, such as the climate crisis, scarcity of raw materials, loss of biodiversity and a more competitive Europe (European Commission, 2021a). The European Union's Green Deal (European Commission, 2021c) and the upcoming taxonomy (European Commission, 2021b) are landmarks in pushing companies from a linear economy (Esposito et al., 2018) to circularity and sustainability. According to the literature review by Kirchherr et al. (2017), "reduce, reuse, recycle (the 3R framework)" (p. 226) is one of the most frequently used frameworks to characterize circular economy. Therefore, circularity aims at maximizing the efficient use of already produced assets (Esposito et al., 2018). Although the narrative around circularity asserts sustainability as an integral part of the concept, often, this is not the case, as sustainability refers to achieving the triple bottom line, i.e., addressing people, planet and profit objectives, *equally* (Jonker et al., 2011). Thus, sustainability aims at a coherent and equally balanced connection between the ecological, economic and the social sphere. However, this contradicts a linear perspective on production and consumption patterns (Schaefer & Crane, 2005). A shift towards circularity and sustainability can then only be achieved by disrupting the traditional way of doing business and hence transforming business models. The key to succeed in this transformation is a company's ability to innovate because circular economy "... requires a fundamental shift instead of incremental twisting of the current system" (Kirchherr et al., 2017, p. 224) and a company's ability to innovate its business model in a responsible way is now the central factor to achieve more sustainability.

While innovation management has focused on creating value with improving conventional products and processes in the past, today's focus is on business model innovation, and a sustainable economic system is becoming the key objective (e.g. Filser et al., 2020; Gassmann et al., 2020; Johnson et al., 2008; Kraus et al., 2020; Lanzolla & Markides, 2021; Prescott & Filatotchev, 2021). Although literature emphasizes the path from a linear to a circular economy and sustainability necessitates new business models (Bocken et al., 2014; Esposito et al., 2018), the majority of business model literature has only marginally or even not at all dealt with circular economy. This has resulted in a separate strand of literature on circular and sustainability-oriented business models (e.g. Antikainen & Valkokari, 2016). Literature reviews (e.g. Bocken et al., 2014; Geissdoerfer et al., 2018, 2020) on circular business models or sustainable business models have not found consensus on the two concepts. Geissdoerfer et al. (2020) define *circular business models* "as business models that are cycling, extending, intensifying, and/or dematerializing material and energy loops to reduce the resource inputs into and the waste and emission leakage out of an organizational system" (p. 7). This perspective does not include any constituent business model aspects (e.g. value proposition or revenue model) but rather refers list of circularity measures and the implementation of such a business model. Bocken et al. (2014), for example, define sustainable business models as

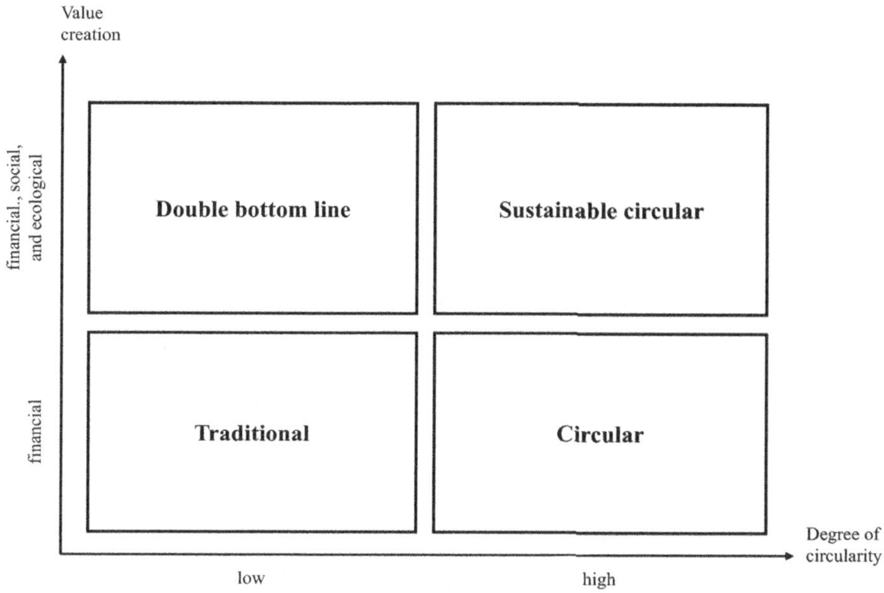

Fig. 1 Archetypes of business models (own illustration)

business models that incorporate "a triple bottom line approach and consider a wide range of stakeholder interests, including environment and society" (p. 42). In addition, Geissdoerfer et al. (2018) offer the following definition of sustainable business models: "as business models that incorporate pro-active multi-stakeholder management, the creation of monetary and non-monetary value for a broad range of stakeholders and hold a long-term perspective" (p. 404). These three definitions exemplify that literature either embraces the essence of sustainability *or* circular economy, not focusing on specific business model characteristics. Further, as stated already by Antikainen and Valkokari (2016), sustainable business models and circular business models are different, yet closely related literature streams with multiple perspectives (Geissdoerfer et al., 2018, 2020; Pieroni et al., 2019). In line with Kirchherr et al. (2017), we therefore conclude that literature has not yet established a clearly defined differentiation *or* link between circularity and sustainability.

To close this gap, we aim to contribute to the question of how circular business models must be enhanced incorporating responsibility and purpose to become truly sustainable. For this, we first propose four archetypes of business models (see Fig. 1) that are existent or emerge. They all have distinctive approaches how and what value to create. In the following section, we illustrate each business model type with examples from the fashion and apparel industry in more detail. Further, we show how companies can transform their respective status quo on different pathways towards sustainable circular business models. We close with a discussion of our findings, the development of three design propositions and the limitations of our study.

2 Theoretical Background: Business Model Frameworks

In this chapter, we structure new business models according to their purpose (financial value protection vs financial, social and ecological value creation) and to the degree to which they integrate circularity (re-use, reduce and recycle) into their business model. The following archetypes result.

2.1 Traditional Business Models

Traditional business models focus on questions about the value proposition and revenue model (Gassmann et al., 2016; Osterwalder & Pigneur, 2010). The success of new and disruptive business models, such as those of Amazon, Airbnb and Uber (e.g. Sorescu, 2017), made this perspective popular, but the traditional business model frameworks equally apply to traditional industries (e.g. Winterhalter et al., 2017). Although the business perspective has offered a more holistic and systemic view in the last 10 years, it lacks the integration of circularity and sustainability. The traditional business model is directly linked to the economic value it creates as the main driver is monetary value for shareholders (i.e. profit or entrepreneur's wage). Thus, these business models make a central contribution to the currently prevailing shareholder economy.

2.2 Circular Business Models

In recent years, companies have started to innovate their business models towards circular economy (Esposito et al., 2018; Geissdoerfer et al., 2020). Although literature has not agreed on a common understanding about the concept (Kirchherr et al., 2017), literature on circular business model innovation shows that organizational (economic) goals, namely, efficiency and effectiveness, are the main objectives. From their literature review, Geissdoerfer et al. (2020) summarize several measures characterizing this type of business model and respective innovations: "recycling measures (closing), efficiency improvements (narrowing), use phase extensions (extending), a more intense use phase (intensifying), and the substitution of product utility by service and software solutions (dematerializing)" (p. 7). Thus, circular business models aim at creating financial value while being triggered by risk mitigation with respect to resource accessibility, financial considerations or regulatory requirements.

2.3 Double Bottom Line Business Models

The raison d'etre of double-bottom-line business models is to address all aspects of the triple bottom line, however strongly focusing on either ecological or social value creation. Double-bottom-line business models are often found in start-ups under, for example, the labels green, social or sustainable enterprises (Gast et al., 2017). Their goal is to generate superior customer value to achieve competitive advantage and capture economic value, as well as to create either a positive ecological or a social impact (Neumeyer & Santos, 2018; Pieroni et al., 2019). They are part of a growing number of companies driven by their purpose and responsibility for multiple stakeholders. The focus of those business models is sustainability, while circularity can be part of it, but it does not have to be.

2.4 Sustainable Circular Business Models

True sustainability encompasses the equal harmonization of the ecological, economic and social spheres—addressing the triple-bottom line (e.g. Weidner et al., 2021). In contrast to business models, which focus on circularity or the double bottom line, this type regards the triple bottom line as an integral part of the concept. Intensifying responsibility and purpose, companies with circular business models will transform into sustainable circular business models, while enterprises with double-line business models strengthen their circularity efforts. The entire value chain of sustainable circular business models is then organized around sustainability and circularity mechanisms, including reducing, reusing, repairing, recycling and responsible disposal. The focus of sustainable circular business models is no longer on creating value for owners, in terms of financial income, but for *all* stakeholders including the planet. With this shift comes a multi-stakeholder perspective (e.g. Freeman, 1984), trying to minimize any harmful impact while simultaneously seeking to create positive effects on society *and* the environment. At the core, sustainable circular business models follow the triple bottom line with responsible leadership and purpose. Additionally, they include circular measures and means into their business model wherever applicable, for example, in their value creation. Those strategies are mainly aiming at minimizing material flows, adopting a resource view.

3 Research Design

3.1 Context of the Study: The Apparel and Footwear Industry

To illustrate differences in current business models and a possible trajectory towards sustainable circular business models, we chose to analyse companies from the apparel and footwear industry for the following reasons: Firstly, this sector is of global importance as it accounts for USD 2.4 trillion in global manufacturing and employs worldwide 86 million people (UN Alliance for Sustainable Fashion, 2021). Second, the fashion industry's business model is currently driven by seasonality of fashion goods, timely fashion trends branding and pricing and cost leadership in production (e.g., Jang et al., 2012). However, the overall understanding suggests that the sector ignores responsibility for human labour conditions and the environment (Joy & Peña, 2017), and thus the business model approach towards sustainability and circularity is strongly limited. The fashion industry is one of the world's most polluting sectors and a significant contributor to climate change and the degradation of the environment (e.g., Berg et al., 2020; Granskog et al., 2020; Joy & Peña, 2017; Niinimäki et al., 2020; Peters et al., 2021). Generally, managerial decisions are guided by speed and price and not by questions of sustainability (e.g. Kedron, 2020). Negative impacts for the environment occur at all stages of the product life cycle (Waste and Resources Action Programme, 2017). The fashion industry emitted about 2.1 billion metric tons of greenhouse gases (GHG) in 2018 alone, counting for about 4% of the global total, equalling the emissions of France, Germany and the United Kingdom combined (Berg et al., 2020). Thus, there is an urgent need to transform the current business model towards sustainability and/or circularity (Andersson et al., 2018).

3.2 Choice of Method and Data

Methodologically, we opt for a secondary data research approach. Illustrating the business models of five companies, we follow the idea of grounded theory (Corbin & Strauss, 2007) with data taken from company websites (Whiteside et al., 2012). Websites are an essential medium in corporate communication strategies (Wheeler & Elkington, 2001) and thus an appropriate source of information for companies' business models (Verboven, 2011). Hence, companies use their websites as a tool to communicate their activities concerning sustainability, circularity and environmental, social and governance (ESG) activities (e.g., Da Giau et al., 2016). From the large number of possible companies, we selected five that are examples of our proposed archetypes. The companies are COS, Levi's, Patagonia, TOMS and Zara (www.cosstores.com, www. Levi.com, www. Patagonia.com, www.toms.com, www.zara.com).

4 Findings: Business Models in the Apparel and Footwear Industry

The following discussion is structured around the business model elements of value proposition (what does the company offer) and value chain (how it is organized to produce the product or service) (e.g. Gassmann et al., 2020).

4.1 The Traditional Business Model of Zara

Zara is the most well-known one brand of the Spanish Inditex Group's eight brands. It is one of the biggest international fashion companies with around 3000 stores (Forbes.com, 2021) and € 20.4 billion in 2020 (INDITEX S.A., 2021). Its business model covers design, production, distribution and sales in an extensive value delivery network. Zara's vision to produce fashion, which is ready-to-wear and comes in regular cycles many times a year, defines Zara's fast fashion approach. Accordingly, Zara's value proposition to design, manufacture and distribute clothes puts the customer at the heart of its business ("Our workforce never loses sight of the customer"). Circularity is seen as a trend that needs to be incorporated into the value creation and delivery process. Thus, Zara's goal is to move in a more circular direction, but the current business model is customer-centred and profit-driven. First measures towards circularity are taken: "By 2023, one of our aims is no longer send anything to landfills from our own headquarters, logistics centres, stores and factories".

4.2 COS and its Circular Business Model

COS is part of H & M, the Swedish fashion giant. It specializes in modern clothing for men and women. Its style is less trend-oriented but inspired by architecture, graphics and design, and it is intended to be worn more than just one season. Global net sales accounted to USD 222.4 million in 2020 (Statista GmbH, 2021). The value proposition of COS is based on a circular understanding of how to create values. This focus also includes actors in the value chain. COS's circular values must be shared by partners from their value network. Value creation is strongly characterized by focal points such as closing material loops, preserving resources, using sustainable materials and extending the life of products. The latter is achieved by providing recommendations for the optimal use of the products. In addition, materials can be re-used through recycling, and COS offers a re-sell shop for pre-worn clothes. COS is currently expanding its measures and has, for instance, set itself the goal to further increase the share of the materials in the collection that were sustainably sourced from already 76% in 2020.

4.3 LEVI's Ecologically Focused Double-Bottom-Line Business Model

LEVI's is a brand of denim jeans from Levi Strauss & Co, an American clothing company headquartered in San Francisco. The iconic brand with its long tradition has approximately 3100 retail stores or shop-in-shops and reported net revenues of USD 4453 million in 2020 (Business Wire, 2021). The value proposition is based on the triple bottom line. However, besides a strong economic focus, the ecological perspective prevails: the brand has introduced innovative steps like a product tag, called "Care Tag for Our Planet", which offers tips for customers on how to best preserve clothing. LEVI's also offers vintage pre-owned or restored items and engages with organizations to collect used clothing in different countries. Thus, the value creation processes are geared towards extending the products' life cycle, preserving resources, closing material loops, re-using materials, saving water and avoiding the use of chemicals in production. The extension of the life cycle is ensured by a repair service. A dedication towards becoming a more sustainable clothing brand can be observed.

4.4 TOMS Shoes: A Double-Bottom-Line Business Model with Social Focus

TOMS shoes was founded in 2006 starting with their first model, One for One®. As a B-Corp, they donate one third of their profits to "grow the good". The respective areas range from building equity at the grassroots level, promoting mental health, increasing access to opportunity to end gun violence and relief during the COVID-19 pandemic. The strong emphasis on social measures, with a dedicated focus on people, is clearly articulated through their mission "... using business to improve lives...". TOMS shoes also create environmental added value being a certified B-Corp, offering Earthwise™ (products rooted in earth-friendly materials and processes and from sustainable cotton). In addition, they have set clear goals towards increased sustainability: "Increase our B Corp environmental score by working to reduce waste and energy use, and by exploring an end-of-use program for our shoes". However, the main characteristic of their business model is their strong focus on creating social value for customers or employees. Strong symbiotic relationships in their value network with partners are proof of this approach.

4.5 The Sustainable Circular Business Model of Patagonia

Patagonia, founded and based in Ventura, California, markets and sells outdoor clothing and gear for climbing, surfing, skiing and snowboarding, fly fishing and trail running. Patagonia counts around 2300 employees in 2021 and reported USD 1 billion for the fiscal year 2017 (craft, 2021). It is well known for its approach to save the planet exemplified by new programs like "Worn Wear" (returned merchandise, which has been cleaned and repaired) and "ReCrafted" (clothing made from scraps of fabric from used Patagonia gear). Patagonia's value proposition is based on a comprehensive triple-bottom-line approach. Elements of this approach are a high focus on compliance with the highest ecological measures and auditing of suppliers through their own sustainability criteria. Premiums are paid directly to the suppliers' factories for compliance with ecological standards. Aspects such as resource preservation, closing material loops, saving water, avoiding pollutants and chemical agents or reusing materials are in the foreground. In addition, there is a high concentration on continuous development, improvement and research of sustainable materials. During the products' use phase, customers receive recommendations for careful use. Patagonia provides a repair and recycling service for the end-of-life phase. There is a strong focus on socio-ecological measures to improve and address socio-ecological problems that go beyond their core business. Patagonia is the only company among those surveyed to openly report negative social and environmental impacts resulting from their value creation and delivery activities.

5 Discussion

5.1 Business Model Innovations and Pathways Towards Sustainability: Three Design Propositions

Literature agrees that business model innovations are necessary on the way to a circular economy. However, academia has not yet found consensus on the concept of business model innovation, its various measures and what exactly constitutes business model innovation towards circularity and/or sustainability (Geissdoerfer et al., 2018). Rather, the two streams of literature are on one side highly interrelated and at the same time understood as two distinct concepts, as reflected in the comparative view by Pieroni et al. (2019), i.e. circular economy vs. sustainability-oriented business model innovations. Literature thus lacks a comprehensive sustainability-circularity concept encompassing economic, ecological and social value creators by means of circular economy.

Based on publicly available information, we found clear differences in how seriously companies take comprehensive sustainability issues, based on which four archetypes were coined (see Fig. 1). We thus suggest that the three-dimensionality of sustainability and measures of circularity should be the guiding principle for business model innovation. However, the trajectory to sustainable circular business models can follow different paths. For this, we develop three design propositions. "A design proposition can be seen as offering a general template for the creation of solutions for a particular class of field problems" (Denyer et al., 2008, p. 395). We

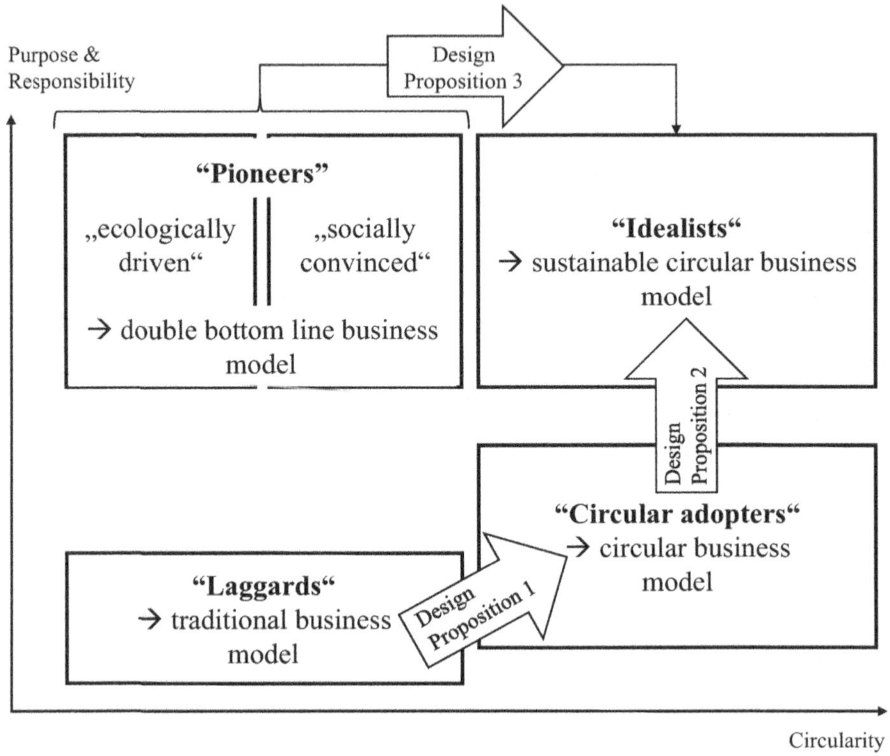

Fig. 2 Business model archetypes and pathways (own illustration)

believe that these design propositions are particularly valuable because in line with suggestions by Denyer et al. (2008), they "contain information on what to do, in which situations, to produce what effect and offer some understanding of why this happens" (p. 396). Thus, our design propositions are guidelines for both academic future field-testing and managers.

Most companies, "the laggards", have business models, which still follow the idea of a linear economy (see Fig. 2). Companies with traditional business models need to re-invent their purpose and start to implement responsible leadership and circular measures all at the same time. Literature shows that behavioural and normative aspects have been neglected in current research on traditional business models as well as on circular business model innovation and/or sustainability-oriented business model innovation (Pieroni et al., 2019). Responsibility is a basic human value and a motivational driver (Schwartz, 1994). As a benevolence value, it aims at preserving and enhancing the welfare of one's in-group and voluntary concern for others. Purpose was originally understood as the functioning of organizations and thus as the maximization of its shareholder's welfare (Gartenberg, 2021). Nowadays, it is increasingly seen as "a company's moral response to its broadly defined responsibilities, not an amoral plan for exploiting commercial opportunity" (Bartlett & Ghoshal, 1994, p. 88). Thus, we suggest that responsibility and purpose are the missing facets in transforming the traditional business model (see Fig. 2). Hence, to innovate a traditional business model's purpose and responsible leadership are prerequisite to implement circularity measures and finally transform the business model towards circularity. This will be accompanied by a change from a shareholder to a multi-stakeholder perspective. This transformation of the majority is particularly valuable to make circular economy mainstream.

Design Proposition 1 *Companies have to revise their purpose, build on responsibility and intensify their measures of circularity companies to transform their business model from a traditional towards a circular one.*

"Circular adopters" have already transformed their traditional business models into circular business models by a more nuanced purpose and increased responsibility (see Fig. 2). However, this archetype is mainly customer- and shareholder-focused following the traditional paradigm of doing business: creating value through sales and profit, measured in monetary units. As exemplified in our cases, COS is an example for these archetypes. Although companies with a circular business model have put sustainability on the agenda, there is no underlying game-changing idea; at best, they put circularity in "the old box" (Hofstetter et al., 2021; Schröder et al., 2019; Vargo, 2021). This approach is rather passive-opportunistic and compliant – in some cases following the notion of anticipatory obedience. To initiate and complete the journey towards a sustainable circular business model, responsibility and purpose are the central levers. "Circular adaptors", still with a limited degree of responsibility and a traditional purpose, must deepen their responsible leadership and align their purpose more strongly towards sustainability to achieve a sustainable circular business model. When responsibility is strongly anchored in the business model, the company can then authentically convey its purpose. We thus propose:

Design Proposition 2 *To transform their business model into a truly sustainable business model,* i.e. *a business model creating economic, social and ecological value by means of circularity in all company operations, companies, with a circular business model, have to intensify their responsibility and purpose as well as their measures and means of circularity.*

"Pioneers", companies with a double-bottom-line business model (either ecologically driven or socially convinced), originally base their business model on high levels of responsibility and purpose towards sustainability with a multi-stakeholder view. Thus, they are active players in the transformation from a linear towards a sustainable economy. They go beyond what is required by law and integrate altruistic behaviour into their businesses. Those approaches are often associated with new forms of corporate governance and management, e.g. holacracy or responsible ownership (Fast Company & Inc, 2019; Wildplastic GmbH, 2021). However, "Pioneers" are not necessarily truly sustainable because circularity is not an absolute condition, and they have not necessarily implemented circularity. Thus, to progress from a sustainable business model to a sustainable circular business model, it is necessary to keep their purpose and responsibility on high level, include one missing aspect of the triple bottom line (either ecological or social) and implement circular measures. Accordingly, we suggest:

Design Proposition 3 *To transform their business model into a truly sustainable business model, companies with a double-bottom-line business model have to keep relying on their responsibility and sustainable purpose, add the missing element of sustainability and intensify their circularity.*

These design propositions extend the work by Bocken et al. (2014) and Geissdoerfer et al. (2018, 2020) and follow Pieroni et al. (2019), who call for an integration and simultaneous application of sustainability and circular business model innovations as sustainability is "more" than circularity and sustainability will not work without circularity. Hence, we go beyond this integration by emphasizing the normative elements (Pieroni et al., 2019) of responsibility and purpose. We suggest an expansion of the current perspective and add to the literature with a comprehensive view on business models. Circularity is a necessary but not sufficient means to increase sustainability and to implement a truly sustainable business model. Accordingly, to focus on either sustainable business models or circular business models only will not bring about the much-needed transformation from a linear to a circular and sustainable economy. Rather, there is an urgent need to integrate the two perspectives.

Responsibility and purpose are managerial efforts that are necessary for this transformation and integration. Responsible leadership and a mind shift from a focus on "administration" (traditional business models) to one on "design" (sustainable circular business models) (Baars, 2018) are the needed changes on the managerial level. While bad designs and restrictive mind-sets offset the transformation towards circular economy and sustainability (Esposito et al., 2018), we call for new ways of doing and thinking. A proven method for grand challenges, respectively, wicked problems, and a promising method to develop sustainable circular business models can be the Design Thinking approach. It is purely human-centred, questions a business's purpose and is driven by responsibility, thus fostering a multi-stakeholder view (Brown, 2009; Dunne & Martin, 2006).

5.2 Limitations and Future Research

Despite our approach, our research is impaired by several limitations. Foremost, our study is limited to one industry, and we base our design propositions on one case per archetype. This limits generalizability per se. The cases are descriptive in nature, and we relied on secondary data from company websites without the inclusion of further secondary sources, such as magazines or social media, as well as stakeholders, such as suppliers. Considering the findings by Cho and Roberts (2010) and the "Toxic 100 list" that shows that the poorest environmental performers make more extensive disclosures on their websites, we indeed acknowledge that this may limit the validity of our conclusions. A second limitation is our approach that only discusses an upgrade in circularity and sustainability. Literature on hybrid organizations (e.g. Ebrahim et al., 2014) has discussed mission drift. If the "Pioneers" with a double-bottom-line business model decrease their responsibility and purpose and do not manage to increase their circular efforts, they will reverse into "Circular Adopters" due to a mission drift. Accordingly, we encourage further research that tackles our limitations and tests our design propositions. Foremost and following our idea that purpose and responsibility are the basis for transforming business models towards true sustainability, we envision future research on the role of responsible leadership or business purpose for business model transformation. Furthermore, we pose the following questions: How do companies tackle the transformation of their business models from a traditional business model to an increased sustainability and circularity? And why do double-bottom-line companies possibly experience or opt for mission drift? Hence, we envision analysis of our proposed business model archetypes in more detail with secondary as well as primary data to find nuanced facets of each archetype. Last but not least, our design propositions and suggestions for futures research should be expanded, for instance, into different manufacturing industries and the service sector (for a review of circularity in tourism see, e.g. Rodríguez et al. (2020)). This may expand our view beyond the standard business focused dimensions to describe business models. This will create new theoretical insights on how companies create and capture value in the current situation where we urgently need to transform the entire economic system towards more sustainability.

References

Andersson, J., Berg, A., Hedrich, S., Ibanez, P., Janmark, J., & Magnus, K.-H. (2018). *Is apparel manufacturing coming home? Nearshoring, automation, and sustainability—establishing a demand-focused apparel value chain*. McKinsey & Company.

Antikainen, M., & Valkokari, K. (2016). A framework for sustainable circular business model innovation. *Technology Innovation Management Review, 6*, 7. http://timreview.ca/article/1000.

Baars, J. E. (2018). *Leading design. Design strategisch einsetzen: Wie Unternehmen das volle Potenzial entfalten!* Vahlen.

Bartlett, C. A., & Ghoshal, S. (1994). Changing the role of top management: Beyond strategy to purpose. *Harvard Business Review, 72*, 79–88.

Berg, A., Magnus, K.-H., Kappelmark, S., Granskog, A., Lee, L., Sawers, C., Polgampola, P., & Lehmann, M. (2020). *Fashion on climate: How the fashion industry can urgently act to reduce its greenhouse gas emissions*. McKinsey & Company and Global Fashion Agenda.

Bocken, N. M. P., Short, S. W., Rana, P., & Evans, S. (2014). A literature and practice review to develop sustainable business model archetypes. *Journal of Cleaner Production, 65*, 42–56. https://doi.org/10.1016/j.jclepro.2013.11.039

Brown, T. (2009). *Change by design: How design thinking transforms organizations and inspires innovation*. HarperCollins.

Business Wire, I. (2021). Levi Strauss & Co. Reports Fourth-Quarter 2020 Financial Results [Online]. Available: https://www.businesswire.com/news/home/20210127005914/en/Levi-Strauss%C2%A0-Co.-Reports-Fourth-Quarter-2020-Financial-Results [Accessed May 12, 2021].

Cho, C. H., & Roberts, R. W. (2010). Environmental reporting on the internet by America's toxic 100: Legitimacy and self-presentation. *International Journal of Accounting Information Systems, 11*, 1–16.

Corbin, J. M., & Strauss, A. C. (2007). *Basics of qualitative research—Techniques and procedures for developing grounded theory*. Sage.

Craft (2021). Patagonia [Online]. Available: https://craft.co/patagonia [Accessed May 12, 2021].

Da Giau, A., Macchion, L., Caniato, F., Caridi, M., Danese, P., Rinaldi, R., & Vinelli, A. (2016). Sustainability practices and web-based communication: An analysis of the Italian fashion industry. *Journal of Fashion Marketing and Management, 20*, 72–88.

Denyer, D., Tranfield, D., & van Aken, J. E. (2008). Developing design propositions through research synthesis. *Organization Studies, 29*(03), 393–413. https://doi.org/10.1177/0170840607088020

Ebrahim, A., Battilana, J., & Mair, J. (2014). The governance of social enterprises: Mission drift and accountability challenges in hybrid organizations. *Research in Organizational Behavior, 34*, 81–100. https://doi.org/10.1016/j.riob.2014.09.001

Dunne, D., & Martin, R. (2006). Design thinking and how it will change management education: An interview and discussion. *Academy of Management Learning & Education, 5*(4), 512–523. https://doi.org/10.5465/amle.2006.23473212

Esposito, M., Tse, T., & Soufani, K. (2018). Introducing a circular economy: New thinking with new managerial and policy implications. *California Management Review, 60*(3), 5–19. https://doi.org/10.1177/0008125618764691

European Commission. (2021a). *EU circular economy action plan: A new circular economy action plan for a cleaner and more competitive Europe*. Retrieved 19.01.2021 from https://ec.europa.eu/environment/circular-economy/

European Commission. (2021b). *EU taxonomy for sustainable activities: Works in view of the establishment of an EU classification system for sustainable activities, i.e. an EU taxonomy*. Retrieved 18.01.2021 from https://ec.europa.eu/info/business-economy-euro/banking-and-finance/sustainable-finance/eu-taxonomy-sustainable-activities_en

European Commission. (2021c). *A European Green Deal: Striving to be the first climate-neutral continent*. Retrieved 18.01.2021 from https://ec.europa.eu/info/strategy/priorities-2019-2024/european-green-deal_en

Fast Company & Inc. (2019). The New Capitalism. Exclusive: Patagonia founder Yvon Chouinard talks about the sustainability myth, the problem with Amazon and why it's not too late to save the planet [Online]. Available: https://www.fastcompany.com/90411397/exclusive-patagonia-founder-yvon-chouinard-talks-about-the-sustainability-myth-the-problem-with-amazon-and-why-its-not-too-late-to-save-the-planet [Accessed April, 30 2021].

Filser, M., Kraus, S., Breier, M., Nenova, I., & Puumalainen, K. (2020). Business model innovation: Identifying foundations and trajectories. *Business Strategy and the Environment, n/a*(n/a). https://doi.org/10.1002/bse.2660

Forbes.com (2021). Zara [Online]. Available: https://www.forbes.com/companies/zara/?sh=5cb838287487 [Accessed May 12, 2021].

Freeman, R. E. (1984). *Strategic management: A stakeholder approach*. Pitman.

Gartenberg, C. M. 2021. Purpose-driven companies and sustainability. Available at SSRN 3786823.

Gassmann, O., Frankenberger, K., & Choudury, M. (2020). *The business model navigator* (2nd ed.). FT Publishing International.

Gassmann, O., Frankenberger, K., & Sauer, R. (2016). *Exploring the field of business model innovation: New theoretical perspectives*. Palgrave Macmillan. https://doi.org/10.1007/978-3-319-41144-6

Gast, J., Gundolf, K., & Cesinger, B. (2017). Doing business in a green way: A systematic review of the ecological sustainability entrepreneurship literature and future research directions. *Journal of Cleaner Production, 147*, 44–56. https://doi.org/10.1016/j.jclepro.2017.01.065

Geissdoerfer, M., Pieroni, M. P. P., Pigosso, D. C. A., & Soufani, K. (2020). Circular business models: A review. *Journal of Cleaner Production, 277*, 123741. https://doi.org/10.1016/j.jclepro.2020.123741

Geissdoerfer, M., Vladimirova, D., & Evans, S. (2018). Sustainable business model innovation: A review. *Journal of Cleaner Production, 198*, 401–416. https://doi.org/10.1016/j.jclepro.2018.06.240

Granskog, A., Laizet, F., Lobis, M., & Sawers, C. (2020). *Biodiversity: The next frontier in sustainable fashion*. McKinsey & Company.

Hofstetter, J. S., De Marchi, V., Sarkis, J., Govindan, K., Klassen, R., Ometto, A. R., Spraul, K. S., Bocken, N., Ashton, W. S., Sharma, S., Jaeger-Erben, M., Jensen, C., Dewick, P., Schröder, P., Sinkovics, N., Ibrahim, S. E., Fiske, L., Goerzen, A. & Vazquez-Brust, D. (2021). From sustainable global value chains to circular economy—Different silos, different perspectives, but many opportunities to build bridges. Circular Economy and Sustainability, n/a.

Inditex S.A. 2021. Inditex: FY2020 results [Online]. Available: https://www.inditex.com/documents/10279/648111/Full+year+2020+Results_v2.pdf/e622f911-a0ee-1103-2b64-0c12829833b5 [Accessed May, 12 2021].

Jang, J., Ko, E., Chun, E., & Lee, E. (2012). A study of a social content model for sustainable development in the fast fashion industry. *Journal of Global Fashion Marketing, 3*, 61–70.

Johnson, M. W., Christensen, C. M., & Kagermann, H. (2008). Reinventing your business model. *Harvard Business Review, December*, 51–59.

Jonker, J., Stark, W., & Tewes, S. (2011). *Corporate social responsibility und nachhaltige Entwicklung: Einführung, Strategie und Glossar*. Springer Gabler.

Joy, A., & Peña, C. (2017). Sustainability and the fashion industry: Conceptualizing nature and traceability. In C. E. Henninger, P. J. Alevizou, H. Goworek, & D. Ryding (Eds.), *Sustainability in fashion: A cradle to upcycle approach*. Springer International Publishing.

Kedron, T. (2020). Cultures of sustainability in the fashion industry. *Fashion Theory, 24*, 715–742.

Kirchherr, J., Reike, D., & Hekkert, M. (2017). Conceptualizing the circular economy: An analysis of 114 definitions. *Resources, Conservation and Recycling, 127*, 221–232. https://doi.org/10.1016/j.resconrec.2017.09.005

Kraus, S., Filser, M., Puumalainen, K., Kailer, N., & Thurner, S. (2020). Business model innovation: A systematic literature review. *International Journal of Innovation and Technology Management, 17*(06), 2050043. https://doi.org/10.1142/s0219877020500431

Lanzolla, G., & Markides, C. (2021). A business model view of strategy. *Journal of Management Studies, 58*(2), 540–553. https://doi.org/10.1111/joms.12580

Neumeyer, X., & Santos, S. C. (2018). Sustainable business models, venture typologies, and entrepreneurial ecosystems: A social network perspective. *Journal of Cleaner Production, 172*, 4565–4579. https://doi.org/10.1016/j.jclepro.2017.08.216

Niinimäki, K., Peters, G., Dahlbo, H., Perry, P., Rissanen, T., & Gwilt, A. (2020). The environmental price of fast fashion. *Nature Reviews Earth & Environment, 1*, 189–200.

Osterwalder, A., & Pigneur, Y. (2010). *Business model generation: A handbook for visionaries, game changers, and challengers*. Wiley.

Peters, G., Li, M., & Lenzen, M. (2021). The need to decelerate fast fashion in a hot climate—A global sustainability perspective on the garment industry. *Journal of Cleaner Production, 295*, 126390.

Pieroni, M. P. P., McAloone, T. C., & Pigosso, D. C. A. (2019). Business model innovation for circular economy and sustainability: A review of approaches. *Journal of Cleaner Production, 215*, 198–216. https://doi.org/10.1016/j.jclepro.2019.01.036

Prescott, J. E., & Filatotchev, I. (2021). The business model phenomenon: Towards theoretical relevance. *Journal of Management Studies, 58*(2), 517–527. https://doi.org/10.1111/joms.12610

Rodríguez, C., Florido, C., & Jacob, M. (2020). Circular economy contributions to the tourism sector: A critical literature review. *Sustainability, 12*, 4338. https://doi.org/10.3390/su12114338

Schaefer, A., & Crane, A. (2005). Addressing sustainability and consumption. *Journal of Macromarketing, 25*(1), 76–92. https://doi.org/10.1177/0276146705274987

Schröder, P., Bengtsson, M., Cohen, M., Dewick, P., Hofstetter, J., & Sarkis, J. (2019). Degrowth within—Aligning circular economy and strong sustainability narratives. *Resources, Conservation and Recycling, 146*, 190–191.

Schwartz, S. H. (1994). Are there universal aspects in the structure and contents of human values? *Journal of Social Issues, 50*(4), 19–45. https://doi.org/10.1111/j.1540-4560.1994.tb01196.x

Sorescu, A. (2017). Data-Driven Business Model Innovation. *Journal of Product Innovation Management, 34*(5), 691–696. https://doi.org/10.1111/jpim.12398

Statista GmbH. (2021). cosstores.com [Online]. Available: https://ecommercedb.com/en/store/cosstores.com [Accessed May 12, 2021].

UN Alliance for Sustainable Fashion. (2021). The clothing and textile industry today. [Online]. Available: https://unfashionalliance.org/ [Accessed March 10, 2021].

Vargo, S. L. 2021. Beyond circularity—A Service-dominant (S-D) logic perspective. Circular Economy and Sustainability, n/a.

Verboven, H. (2011). Communicating CSR and Business identity in the chemical industry through mission slogans. *Business Communication Quarterly, 74*, 415–431.

Waste and Resources Action Programme. (2017). *Valuing our clothes: The cost of UK fashion*. WRAP.

Weidner, K., Nakata, C., & Zhu, Z. (2021). Sustainable innovation and the triple bottom-line: A market-based capabilities and stakeholder perspective. *Journal of Marketing Theory and Practice, 29*(2), 141–161. https://doi.org/10.1080/10696679.2020.1798253

Wheeler, D., & Elkington, J. (2001). The end of the corporate environmental report? Or the advent of cybernetic sustainability reporting and communication. *Business Strategy and the Environment, 10*, 1–14.

Whiteside, M., Mills, J., & McCalman, J. (2012). Using secondary data for grounded theory analysis. *Australian Social Work, 65*, 504–516.

Wildplastic GmbH. (2021). Wildbag [Online]. Available: https://www.wildplastic.com/en/ [Accessed May 10, 2021].

Winterhalter, S., Weiblen, T., Wecht Christoph, H., & Gassmann, O. (2017). Business model innovation processes in large corporations: Insights from BASF. *Journal of Business Strategy, 38*(2), 62–75. https://doi.org/10.1108/JBS-10-2016-0116

Lean Startup and Sustainable Business Model Innovation: A Review of the Customer Development Process

Alberto Peralta

Abstract Since its release, the Lean Startup (LS) method has taken the business model innovation world like a storm. Subject to praise and rejection by many, this practice is now mainstream for incubators, accelerators, and innovation centers. Academics are devoting many research hours to work out LS implications. Yet we found that LS application in the field of sustainable business model innovation (SBMI) has been overlooked. Thus, this chapter concentrates on deepening on three specific LS features: a set of 17 claims called the Customer Development Manifesto (CDM), the core LS process called the Customer Development (CusDev), and the suitability of the CDM and CusDev for SBMI. A thorough review of the academic innovation literature and fields has helped us rigorously examine the CDM claims finding sound supportive foundations and limiting considerations—serving as future research avenues. Extending the implications of the CDM into the CusDev practice, we also found academic support to base its approach to modelling businesses and incorporating stakeholders to this modelling. Moreover, having defined the valueholder concept as a subset of relevant (salient and fringe) stakeholders, we have widened CusDev original design to properly shape sustainable business models integrating valueholders' interests with the activities and challenges imposed by SBMI and its triple (economic, social, and environmental) bottom line. The chapter closes with implications for researchers and practitioners and future research proposals.

Supplementary Information The online version contains supplementary material available at [https://doi.org/10.1007/978-3-031-08313-6_4].

A. Peralta (✉)
Faculty of Economics and Business Administration, University of Alcala, Alcala de Henares, Spain

College of Business, Abu Dhabi University, Abu Dhabi, UAE
e-mail: Alberto.Peralta@edu.uah.es

V. Prokop et al. (eds.), *Business Models for the Circular Economy*, Sustainability and Innovation, https://doi.org/10.1007/978-3-031-08313-6_4

1 Introduction

The book *The Lean Startup* by Eric Ries (Ries, 2011) has received a lot of attention due to its practice orientation and seeming soundness. Given the failure rates of new product development (NPD)-based endeavors (Nightingale & Coad, 2014), it becomes no surprise that entrepreneurs, corporations, and governments aim for other ways to develop businesses. Nevertheless, very few have reported alternatives to new business development departing from one idea and following a process like the NPD, i.e., the linear stage-gate process that starts with idea generation and ends with post-sales services. To this point, the introduction of the Lean Startup (LS) in 2008 was understood by many as an alternative process that replaces the stage-gate process by the Customer Development (CusDev) process, i.e., the construction of a business model (BM) that can grow. Therefore, LS aims to innovate and grow BMs, rather than ideas. Over the years, LS has turned out well-known by some other complementary features (Frederiksen & Brem, 2017): user and customer involvement in product and business development, an iterative approach to new product development, experimentation in new product development, the minimum viable product, and entrepreneurial thinking and planning.

The stronger attention of LS to BMs and stakeholders (Blank & Newell, 2017; York & Danes, 2014) underscores, we argue, its improved effectiveness as a sustainable business model innovation (SBMI) practice than the conventional NPD. SBMI describes the type of innovations that "reduce the environmental impact caused by consumption and production activities" (Carrillo-Hermosilla et al., 2010). When applied to BM innovation (Cavalcante et al., 2011; Chesbrough, 2010; Christensen et al., 2016), SBMI promotes new or updated BMs that comply with their stakeholders (Donaldson & Preston, 1995; Freeman et al., 2004; Mitchell et al., 1997) and their economic, social, and environmental stakes, needs, and interests (Elkington, 2004; Schaltegger et al., 2012). More specifically, sustainable business models (SBMs; Nancy Bocken et al., 2013; Stubbs & Cocklin, 2008) aim to satisfy the interests of their valueholders, i.e., a crucial stakeholders subset for the immediate survival and growth of a new BM (Peralta et al., 2018). Consequently, we contend that LS-based BMs emerge in a much better shape to address valueholders and leverage their interests to meet sustainability ambitions than NPD-based BMs.

LS implications are thus many and large stemming from its core process called the Customer Development (CusDev) and its core claims called the Customer Development Manifesto (CDM: Blank, 2007; Blank & Dorf, 2012). Nevertheless, some authors (e.g., Frederiksen & Brem, 2017) featured those implications from a scientific perspective. Others, like Eisenmann et al. (2011), York and Danes (2014), or Täuscher and Abdelkafi (2016) critically presented what they think was relevant about LS. Yet, most researchers have missed LS core underpinnings represented by the CusDev and the CDM. These were laid out by Blank (2013) and Blank and Dorf (2012) as a four-stage cyclic process and a 17-principle list alluding to strategy, new-product process, engineering, organization, financial reporting, failure, and speed.

Given LS relevance in today's incubators, accelerators, funding rounds, entrepreneurship courses, and the development of experiment strategies to lower new BMs' failure rate, we feel that research on LS as a tool for SBMI remains incomplete from two stances. One, the CDM stays understudied, its theoretical appropriateness, implications for current research and practice, and impact on SBMs sustainability approach. Second, CusDev persists overlooked by LS researchers, and its stages and prescriptions to fill SBM stakeholders and valueholders' needs deserve an in-depth exploration. In this chapter, we ambition to deepen the knowledge on both streams answering three questions: what implications has the CDM for SBMs sustainability goals? How do SBMs react to valueholders? How does CusDev encourage SBMI?

The paper is relevant for entrepreneurs, innovators, analysts, governors, and researchers willing to approach LS as a sound practice to achieve SBMs balanced triple bottom line—balanced economic, social, and environmental goals. SBMs, and SBMI, demand today more comprehensive methods and practices to incorporate stakeholders in their strategies, implementations, and growth. But working with stakeholders following NPD's linear and top-down approach only barely addresses the multiplicity of features linked to each and all of them. LS shows an enhanced ability to work with multiple stakeholders, even with valueholders, owing to its bottom-up and experimentation approach.

After this introduction, we lay out our mix-methods approach including academic publications and a selected set of cases. We review the academic support and discrepancies of the CDM comparing it with different frameworks and provide a sound ground for its principles. We continue describing CusDev and how SBMs might benefit from this practice. Finally, we draw our conclusions and derive theoretical and practical implications from our investigation.

2 Method

2.1 Approach to Assess the Customer Development Manifesto

In this conceptual article, we review the Customer Development Manifesto (CDM), the Customer Development process (CusDev), and the valueholder notion to answer deepen on their implications for sustainable business models (SBMs). Therefore, our approach starts with the analysis of Blank and Dorf self-recognized sources (shown verbatim from 2012: 551): Lead User Research (Von Hippel), Crossing the Chasm (Moore), Entrepreneurial Mindset (McGrath/MacMillan), Innovators Dilemma (Christensen), Profitable Value (Lanning), Lanchester Strategy (Yano), High-Tech Marketing (Davidow), OODA Loop (Boyd), Question-based Selling (Freese), Solution Selling (Bosworth), Conceptual/Strategic Selling (Heiman), Spin Selling (Rackham), US Marine Corps Warfighting Manual, and Tipping Point (Gladwell)—although far from any academic referencing style, these references provided a valuable starting point for our literature search.

The exploratory research to answer our first research question on the CDM started using a back-and-forth reference process to find evidence for or against the CDM statements. We searched and handpicked the references according to each of the CDM's statements and descriptions in Blank & Dorf, 2012, and their subsequent writings. Using this guide, we selected the search strings and searched various databases (Web of Science, ABI/Inform ProQuest, and Google Scholar if needed). (See Data 1, Table A1 in the Appendix 1 for the set of references we used in this chapter).

Moreover, we have also looked at each of the statements of the CDM from an eco-perspective trying to answer the question of whether the CDM would guide the LS method to innovate SBMs. For the connection of the CDM with sustainable business model innovation (SBMI), we decided to use a similar review approach. In this case, none of the extant SBMI literature links with the CDM. Then, our task here was to deduct whether the CDM claims made any difference from a selection of peer-review SBMI literature. This selection followed the same rationale explained earlier. (Refer to Data 1, Table A2 in Appendix 1 for the eco-innovation references we used in the paper).

2.2 *Approach to Assess the Customer Development Process*

To answer our second and third research questions and validate CusDev approach to valueholders and its relevance for SBMI, we firstly searched WOS and ABI/Inform ProQuest databases for peer-review papers in English containing the string "Lean Startup" in their titles or abstracts, from 2012 to date. We chose this search string because the "customer development" string brought no results. Given the authors' practice experience with the LS method, the explicit connection of LS and CusDev (Blank, 2013; Ries, 2009, 2011), and the in-depth analysis of each of the papers identified with the "lean startup" string, the review process brought up the evidences needed to answer the research question.

After an initial selection of 46 papers matching the first criterion, we selected those devoted to properly explaining the CusDev method and its practice and showed connections between it and sustainable innovation concepts, reducing the number of papers to nine. Out of these nine, none of them used the term "eco-innovation," "green innovation," "environmental innovation," or "sustainable innovation" (Boons & Lüdeke-Freund, 2013), but an in-depth analysis of each article allowed us extract the conceptual relationships between LS and sustainable innovation.

Similarly, we conducted a literature review on the same databases using the strings "business model innovation" *and* "sustainable or sustainability" and "business model" *and* "eco-innovation" for the period 2014 to date. Out of the resulting 64 papers, 14 dealt with how BMs are sustainably innovated. They provided us with a deep understanding of the concepts, relationships, and key elements that we needed to frame the conceptual connection between CusDev and SBMI.

We then presented our evidence to three SBMI experts and five LS experts, who helped us assess the literature references with their opinions and new references. At length, we gathered 13 references helping us validate CusDev as a SBMI method (See Data 1, Table A3 in the Appendix 1).

3 The Customer Development Process under Review

3.1 *Building Support for the Customer Development: Principles and Claims of the Customer Development Manifesto*

Blank and Dorf (2012) originally modelled a set of 14 principles that they called the Customer Development Manifesto (CDM). Their aim was to lay out the "rules for startups and big companies alike when they commit to the customer development process" (Dorf, 2013). (See Data 1, Table A4 in Appendix 1 for a comparison of the original organization of the principles (Blank & Dorf, 2012) and a later version (Blank, 2012)). Following Blank (2012), the CDM 17 claims plus a guiding principle are notably evident; the updated version might respond to Blank's intentions to ease the understanding of his principles of entrepreneurship and business model innovation. We believe that the main contribution of the CDM updated version is the guiding principle, which was part of a full chapter in Blank and Dorf (2012: 1–18) but was left out of the CDM claims. For matters of this chapter, we will use the latest 2012 version.

3.2 *Analysis of the Claims of the Customer Development Manifesto*

As a manifesto,[1] the CDM states the motives of the CusDev process departing from one general principle: "A startup is a temporary organization designed to search for a repeatable and scalable business model" (Blank & Dorf, 2012, pxvii). This is to say that a startup emerges as the construct of innovators to innovate business models (BMs); and this construct is short-lived and must quickly search for very special types of BMs.

This definition of a startup contends with certain literature that distinguishes (early stage) startups from (established businesses) incumbents (Rissanen & Sainio, 2016; Picken, 2017), when referring to business model innovation and how the two types of entities approach this innovation. For this literature, the two types of entities

[1] "Manifesto." Merriam-Webster.com Dictionary, Merriam-Webster, https://www.merriam-webster.com/dictionary/manifesto. Accessed 29 Nov. 2021

seem evolution stages of a company. On the contrary, the CDM opens the startup concept, in line with (Christensen et al., 2016), describing a business strategy that deals with the uncertainty and complexity of BM innovation, independent of entrepreneurs or C-level executives.

In our opinion, the most interesting aspects of the CDM definitions are its strategic claims and their implications for new business modelling. "Search" becomes the strategic keyword. Whether changing an existing BM or creating a disruptive one (Khanagha et al., 2014), when innovating a BM, the CDM recommended strategy is the "search" for that BM. The search process must be executed through experimentation or validation cycles of the new BM *elements*. As an example, are "operational excellence," "customer intimacy," or "product leadership" (Treacy & Wiersema, 1993) where developers of the new BM should concentrate on? Or is it a combination of the three? The answer, if any, must be found through the "search" strategy. The CusDev process becomes then the guide to design and implement this strategy.

The proverbial "search" strategy goal, i.e., a repeatable and scalable BM, is also clearly stated in the CDM. In this case, the CDM explicitly describes the goal beyond the creation of any BM—a construct that creates, delivers, and captures value (Osterwalder, 2004; D. J. Teece, 2010). The search must find at least one BM that can repeat sales ("repeatable"; (Blank, 2007) and can grow ("scale"; ditto); these two requirements broadly define success (Powell & Bitner, 1992) of a new BM for the CDM and help set the success path (Picken, 2017 Fig. 1, p588).

In Table 1, we analyze each of the CDM claims, considering the support found in the literature selection (see Data 1, Appendix 2 for the full analysis of each CDM claim).

3.3 Overview of the Customer Development Process

The Customer Development Manifesto (CDM), as a list of guiding claims, leads the Customer Development (CusDev) process (Blank, 2007). This process is the method that Blank developed to "reexamine the existing product-introduction process" and to "delineate a radically different method that brings customers and their needs headfirst into the process long before the launch" (Blank & Dorf, 2012: p xiv). After its first introduction (Blank, 2007), the process has been further refined and enhanced by its use by thousands of entrepreneurs, corporate innovators, and institutions. Although this chapter refrains from assessing CusDev success, or its validity, leaving these analyses for future investigations, it should be noticed that there are many reports about its worldwide use (as a hint, the book by Blank and Dorf has been translated into more than nine languages). This wide acceptance of the method is the reason behind our interest in the principles that guide, and the sustainable impact of, CusDev.

In our opinion, the deepest redefinition of the CusDev process came with the release of Ries' "The lean startup method" (2011). This CusDev latest version

Table 1 Summary of evidence for and against the Customer Development Manifesto claims (Developed from Blank, 2012; Blank & Dorf, 2012)

CDM rules	Evidence for	Evidence against	Remarks
Rule no. 1: "There are no facts inside your building, so get outside"	Cassar and Craig (2009) and others show evidences the tendency of founders to overstate their chances of success at starting up, even seasoned entrepreneurs or founders This claim is in line with Von Hippel (1986) and Sarasvathy (2001). From different perspectives ("lead users" and "effectuation"), they acknowledge that conventional market research to build/renovate new products, or "artifacts," cannot follow the same causal process as in situations of preexistent goals, companies, markets, or economies	Although we have not found evidence contrary to these claims, Von Hippel and others seem to suggest that these findings and feedback points should be somehow digested by the analysts' abilities and then passed on to the decision makers. Then, from a normative theory of choice stand, where premises are set, they may contradict this CDM claim	Probably, the inbound flows with other stakeholders different from customers remain to be assessed. The experimentation regarded in the literature only seems to take customers into account. The CDM overly concentrates on valueholders New lines of research will test the innovation processes when goals and value premises are fully determined (conventional assumption of BMI) and their differences when those premises are inexistent, or barely defined (the CDM assumption). Further, how to treat both circumstances needs exploration .
Rule no. 2: "Pair customer development with agile development"	Agile development evolved from its origins as an alternative software development practice to a plan-based, sequential, predictable development. Iterative and incremental (agile) software development, on the contrary, is creative and people-centered, caring less for the process and more for quick response (a comprehensive analysis of agile and its foundations can be found in Dybå & Dingsøyr, 2008, and Tripp et al., 2018)	Whether sequential (waterfall) or agile, engineering models do not usually consider any forms of value beyond compliance with user specifications. They disconnect those specifications from other stakeholders, or valueholders (Dennehy et al., 2016; Baumeister et al., 2017, pp. 37–51). And they certainly detach value creation, form delivery and capture, or exploration for uncaptured value (Teece, 2010; Yang et al., 2017). Recently, we have	Our literature review suggests that authors regard hardware/software separation as an "either-or" type of situation. Yet, reality shows that today's split between software and hardware is narrowing, even inexistent. The CDM likely acknowledges this circumstance, and we do in a similar way in Fig. 3 When innovating a BM, the process usually produces several new BMs—Some become ephemeral. Further investigations on this

(continued)

Table 1 (continued)

CDM rules	Evidence for	Evidence against	Remarks
		witnessed the emergence of Requirements Engineering, combining agile with Human-Centered Design, but certainly we find much struggle yet to be solved in this domain (Baumeister et al., 2017, pp. 37–51)	multilinear, multi-development conduct are needed
Rule no. 3, "Failure is an integral part of the search for the BM," and Rule no. 4, "If you're afraid to fail you're destined to do so"	Based on raw data and popular literature (Mullins & Komisar, 2009; Sebastian Quintero, 2017), we know the odds for success of new BMs are few (Giardino et al., 2014; Mullins & Komisar, 2009; Unterkalmsteiner et al., 2016). And the odds affect new and established BMs. The two BM types have differences related to communication, coordination, client history, resources, and influences (Sutton, 2000), which make failures being regarded quite differently (Giardino et al., 2014) on each case	Block and MacMillan (1985) argue that assumptions should not define timings, whereas the CDM imply that decisive actions should be taken with speed They also advocate for a linear, milestone type of path, reflecting the product maturity not the BM maturity Control of experiments and results is guided by four different dimensions: team, product, business, and market (Giardino et al., 2014)	Here, the CDM does not advocate for a "failure" strategy. The CDM raises the attention to integrate failure when starting a new BM We have found many conceptual and factual evidence supporting this claim. The CDM builds on failures from an iterative, evolutionary learning perspective based on multiple and rapid testing New research might concentrate on addressing validation tests (design, application, quantity, timing) while innovating BMs, the effect of/on teams (Unterkalmsteiner et al., 2016), and on finding ways of using failure conclusions into new developments
Rule no. 5: "Iterations and pivots are driven by insight"	CusDev is fundamentally about building the micro-foundations that sustain the organizational level variables affecting all elements of the BM (Crossan & Apaydin, 2010; Felin & Foss, 2005). It is an orderly process	Echoing Felin and Foss (2005); p. 450), the CDM and related CusDev leave questions like "who starts the firm, who is attracted into it, who turns over, who the organization is composed of, etc." half-answered	Pivots point to people, to individuals making decisions. The CDM addresses those individuals whose characteristics and conditions influence the (back and forth) progress of the new BM. Further research is

	to generate and evolve the routines and dynamic capabilities that later will affect firm's performance (Felin & Foss, 2005). The process starts from "the purposeful and intentional action" of individuals, i.e., valueholders (e.g., founders, shareholders) that impose pivots and iterations. Valueholders expectations, empowered and prioritized, will form the "organizational code" (Felin & Foss, 2005) and produce "organizational debt" as well as "technical" and financial debts (Blank, 2015b)	at best. The BM innovation process is probably at the root of the heterogeneity of firm performance, but probably so because of the "initial conditions, decisions, and even the characteristics" of the individuals (stakeholders) involved in this process. The CDM seldom treats these conditions, assuming some sort of random homogeneity of stakeholders (founders, investors, early employees, customers, and environment conditions, to name some) and partially avoiding the *who?* question (Felin & Foss, 2005, p. 450)	needed in how decision-taking individuals require main and supporting BMs and which implications the pivots have in the development of subsequent BMs
Rule no. 6: "Validate your hypotheses with experiments"	The CDM rests heavily on experiential, rather than analytical, learning that tests combinations of value propositions, resources, and processes to deliver and capture the right value for the firm and its valueholders (Lanning, 1998; O'Connor et al., 2008). One proposed question to develop the right BM is how to make and sell what product? Experiential learning answers this question focusing on managing the experiences the intended customers, and the rest of the valueholders, have	The exact ways the CDM proposes how this task is accomplish are unclear. It states that the learning needs to be extracted from the customers, but not under which conditions (O'Connor et al., 2008), nor contexts (incremental or radical innovation; Forés & Camisón, 2016). CusDev neither elaborates on other valueholders besides customers as sources of insights	It remains to be understood how the CDM addresses different market paces, or the degree of knowledge needed about a market, as they may be determinants of a more exploratory learning method, or a more assessing method, and which techniques to use Skills are required to (1) design experiments so that learning is maximized (Sitkin, 1992); (2) conduct experiments with goals in mind as opposed to processes; (3) evaluate results given the highly ambiguous action: outcome relationship; and (4) decide on the next course of action based upon strategic

(continued)

Table 1 (continued)

CDM rules	Evidence for	Evidence against	Remarks
			objectives. Iteration is critical as new information and changing conditions are uncovered
Rule no. 7: "Success begins with buy-in from investors and co-founders"	Out of the principles of the CDM, only Rules no. 7 and no. 11 are focused on specific elements of the BM: No. 7 in founders, investors, and early employees and no. 11 in part of the value proposal	The "learning and discovery" approach of the CDM makes it very difficult to a priori comply with Zott, Amit, and Massa (2011: p. 10) request for analysis of the connection of other BM elements and other constructs, because elements and constructs need to be (1) discovered, and (2) so do their relationships. And the stream of claims within Rule No.7 clearly hypothesizes around what (the philosophy) the connection among these elements needs to be	Although it is recognized that the most important information for a firm is the "information flowing in from the market" (Leonard-Barton, 1995), p177); (Yuan et al., 2014) further research is needed on the effects that a radically different method like CusDev produces in an organization leading team. This most probably will be related to a cognitive framework change that will force top teams to explicitly differentiate managing conventional innovation functions and the CDM-based innovation functions
Rule no. 8: "No business plan survives first contact with customers"	Following Moltke's quote, "no plan of operations extends with any certainty beyond the first contact with the main hostile force" (von Moltke, 1871). Or as Mintzberg stated it, "strategic planning is not strategic thinking" (Mintzberg, 1994). So the CDM states the need for other tools to form new BMs (McGrath, 2010: 252, 256) The BM concept fosters innovative thinking. It also allows	Calculated strategies, as opposed to commitment management, is the conventional way of acting. Since Porter, the conventional wisdom, says: "first an objective should be established." And then calculations are made to get there (Mintzberg, 1994). "Coherent strategy planning and execution" is a common theorized linkage between strategy and firm performance that is also used for connecting BM	Many established companies find BM innovation economically unrewarding (McGrath, 2010, showed the reasons), which means that Rule No. 8 claims make sense only in the exception of a firm willing to create BM innovations. For those firms and founders, further research is needed to understand how the BM design (based on discovery-driven planning and experimentation) is superior to the

	multidimensionality, which programming or business planning hinders or spoils (Garvin et al., 2008)	innovation and firm performance (Bengtsson & Tavassoli, 2018). But "coherence" is not resolved there, as it may refer to an integration of BMI planning and execution more than to the use of any other tactic	business plan when a firm (1) enters a new market with entrenched competitors or (2) faces a crisis of their core BM or (3) attempts to scale up a new-to-the-world product (C. Markides, 2006). Also, new research is required to understand how and when to replace conventional strategic planning (based on the conventional business plan) by discovery-driven planning when venturing into new BMs
Rule no. 9, "Not all startups are alike," and Rule no. 11, "Agree on market type—It changes everything"	Arguably, the startup categorization resembles an update of Ansoff's product-market strategies (1957). It balances technology push and market pull, keeping Ansoff's categories or configurations (Verhoeven & Johnson, 2017). But it also covers the realization of the new relation product market that we have experienced with the advent of the democratization of production and distribution means, and the ease of customers' search for solutions, characterized by the "long-tail" concept (Anderson, 2006; Rieple & Pisano, 2015)	Although given the practice-based departing point of the CDM, and assuming a certain misinterpretation risk from our side, it could be argued that these CDM categories form a conceptual typology of startups (BMs). The ontology may not be accurate in reflecting the reality and certainly has limited explanatory power (Harms et al., 2007). Given the lack of evidence for or against its case, but assuming the importance of the market-type factor, recommendations are made to sustain these claims from empirical and theoretical perspectives (Sharma, 2004)	Empirical evidence of the startup classification, its importance for survival, and the effect of changes from one market type to another due to growth is needed. In this regard, and to avoid linear cross-sectional/contingency studies, or the use of improperly scaled unobserved variables, Harms et al. (2007: 664) suggest the application of the configuration approach. This approach would help relate market type with other variables, described also by the CDM. In Harms et al. classification, this type of research fits as a third stage, i.e., one that addresses the evolution of the startup since inception to maturity
	From a supply-side perspective, innovation metrics may serve two	Davila et al. (2007) found that poor or inexistence of management control	The CDM emphasis on metrics that help control the new BM progress

(continued)

Table 1 (continued)

CDM rules	Evidence for	Evidence against	Remarks
Rule no. 10: "Startup metrics are different from those in existing companies"	purposes: (1) assessing the performance (effectiveness and efficiency; Rompho, 2018) of the new BM vs. a comparative forecast or established model and (2) validating the adequacy of the new BM to integrate with existing models and serve valueholders The (1) purpose is usually addressed by NPD metrics although NPD perspective might be limited, as the CDM and Muller et al. (2005) denunciate For the (2) purpose, the new BM poses risks of cannibalization of existing models and even aversion to other risks associated with the new BM. Thus, innovation metrics, rather than conventional metrics, help define behaviors aligning actions with goals (Muller et al., 2005)	systems at startup companies is negatively associated with growth and positively linked to higher turnover of CEOs of startup companies. They adopted a research perspective based on existing business functions (finance, human resource, planning, product, sales/marketing, and partnerships) which seems contrary to the CDM claims. Although their findings may be inconclusive for early-stage startups, it seems the early adoption of financial planning (operating budgets, cash flow, and sales projections) and evaluation systems (capital investment, operating expenses, performance against financial targets) might help startups while growing (Davila et al., 2007). But they acknowledge differences in the application of those tools related to the followed strategy type (cost vs. differentiation)	from the demand and supply sides turns uncommon. Further investigations should address this broad framework, and confirm which metrics are relevant to grow and achieve success Equally important, we recognize the need of making these metrics relate to speed and timing. Research is still infant in this direction And finally, the multilinear reality of business modelling remains unaddressed until now. Figs. 2 and 3 present a configuration for the development of those models that compel complex, non-linear trajectories
Rule no. 12: "Fast, fearless decision-making, cycle time, speed, and tempo"	These demand-side claims make explicit the need to adapt to the fast cycle times of most of today's markets The odds of finding the right BM increase with increased relationships with key customers (Trimi &	Without disregarding the CDM claims, Tornikoski and Renko (2014) advocate for the connection between success and the type of opportunities that startups pursue ("organizational emergence," p. 193). They are the main factors affecting the creation	Very little is researched on the impact of speed in the success of new BMs. Practices like shortening cycle times or speeding the expansion of relationships with key customers seem standard across industries. But we lack knowledge

	Berbegal-Mirabent, 2012). These interactions/tests should be iterative and incremental, with flexible improvements of the model (agile view, (Beck, 2001; Trimi & Berbegal-Mirabent, 2012). But they also should minimize waste, shorten cycles, and increase cycle times (Trimi & Berbegal-Mirabent, 2012) .	speed of a new BM/startup. Other factors affect the startup growth process as well: industry or founding conditions are two additional examples	about the working mechanisms between the internal and external conditions and the speed of BM development. And we have insufficient knowledge about how that relationship moderates/mediates successful BM development Additionally, we must understand better the differentiation between speed outside and inside the startup/BM, how they matter, and to which extent
Rule no. 13: "If it's not about passion, you're dead the day you opened your doors"	The evidences that relate commitment (passion and drive) with successful innovation turn out well supported (Jain, 2011; Sai Manohar & Pandit, 2013). It is by "walking their talk" and by "commonly practicing (by everyone) the core values" that successful innovation (product, process, or BM) takes place Additionally, these claims may be supported by the conceptualization of the *adaptation* and *improvisation* constructs (Duxbury, 2014). As noted by the CDM and authors like Duxbury (2014), founders and BM innovators face turbulences, change, ambiguity, and uncertainty, and they usually respond with rapid learning/experimentation, creativity, and action	Although speed and uncertainty are norm when creating a new BM, the CDM addresses one big strategy to cope with them. Successful entrepreneurs master behavioral traits like *adaptation* and *improvisation*, coupled with execution at the right time. For example, most likely, adaptation will be a safe behavioral strategy, yet customer needs research does not always require fast actions. In addition, improvisation, as a spontaneous conduct, under situations of no-referents, time-pressed conditions, and complex, multilinear paths, requires certain attention	These CDM claims help understand what probably may be the "necessary" (not sufficient, as other psychological, social, and situational factors and functions have to be present; Jain, 2011) success factors of a new BM and its growth into a full-scale company Still, the CDM refers to the individual traits. And we need further exploration of how *commitment* ("passion and drive"), *adaptation*, and *improvisation* play their part in the formation of the core values of the future organization. Surely, further assessment to calibrate which behavior fits best and how to train them is needed

(continued)

Table 1 (continued)

CDM rules	Evidence for	Evidence against	Remarks
Rule no. 14, "Startup titles and functions are very different from a company's," and Rule no. 17, "Startups demand comfort with chaos and uncertainty"	It has been profusely grounded that startups and new BMs require the lead of a particular type of individuals (e.g., Crossan & Apaydin, 2010 1170: Cueto & Zhang, 2014: p. 4). And a similar description to these CDM claims is extensively found in the effectuation theory (Sarasvathy, 2001; Ye, 2016) Founders of new businesses emerge from actual losses, alliances, and leveraging of contingencies (or opportunities). Their constant revision of goals and actions defines the individuals that intend to control the future, usually based on heuristics, rather than predict it. As such, their logic differentiates the entrepreneurial mindset from a more managerial one (Sarasvathy, 2001; Ye, 2016)	The debate between the discovery (realist), the creation (effectual), the complexity, and the Austrian schools of understanding opportunities and the origins of a new BM is far from the end (Phelan, 2015). This struggle helps us further develop the individual profile of entrepreneurs/innovators. Specifically, the debate centers at how innovators' individual profile and behavior might be influenced by the balance between indicators of effectiveness (causal) (Powell & Bitner, 1992) and indicators of process control (effectual) (Makridakis et al., 2010; Sarasvathy, 2001)	One cited consequence of the effectual way of entrepreneurs and innovators was that they used less information and formal analysis techniques than the more causal managers (Ye, 2016: p. 403). They might be the type of people that trust simple statistical and control tools rather than sophisticated algorithms (Makridakis et al., 2010). We need further refinements of the effect of these simpler techniques, and their lack of predictive behavior in favor of more control, as the CDM rules claim Are all successful entrepreneurs effectual, at all times? Are opportunities created, searchable, and plottable, forming a "dancing landscape" (Phelan, 2015)? How broad a range of possibilities might be to fit 95%, or 90% accuracy? We need more investigations about these questions
Rule no. 15: "Preserve cash while searching. After it's found, spend"	Financial management is recognized as "the center of the overall management system" (Karadag, 2015; López et al., 2012). This bold statement undoubtedly points at the effect of the inefficiencies in survival and	Strategic management (Karadag, 2015: p. 28) may result controversial for new BMs. Much of its tenets and requirements seem inherited from larger, established companies (see López et al., 2012: p. 102). Although	Financial strategic management is frequently associated with causal, discovery planning. In those situations, where uncertainty might be represented by a set of heuristics, and where planning and forecasting

	performance of startups and new BMs if poor or lack of financial management is present in them (Karadag, 2015) Although insufficient for causal authors (Karadag, 2015), cash management seems universally recognized as the reason for most business failures, coupled with receivables and a more working capital management (Karadag, 2015: p. 33)	much is said about the differences in uncertainty or the entrepreneur/manager different approaches (to financial management), academics still operationally neglect them in the strategic management field and in the financial strategic field. In other words, much of the lack of practices associated with poor and secondary financial strategies in startups and new BMs is related to the inapplicability of financial precepts and financial literacy in the innovation context (López et al., 2012)—and the prevalence of the operations or marketing/sales functions (Karadag, 2015)	have the support of an unearthed probability function, we know that financial management models like Karadag's (2015) help new BMs focus on what's important, i.e., the correct preservation of cash But causal planning seems ill suited for less controllable uncertainty. Unexpected, *coconut* uncertainty (Makridakis et al., 2010) requires a more effectual financial management. Studies on how to effectively set the financial strategy while creating a new BM and the relative/absolute importance of the financial strategy in the overall context of that creation are in their infancy, and further investigations are certainly needed
Rule no. 16: "Communicate and share learning"	When starting up, a new BM founding team faces informal communication and a loose organizational structure, and eventually everything should be fluid (Picken, 2017). But as the BM progresses and scales, structure, processes, and discipline gain in, providing effectiveness and specialization. In some instances, this final stage of the transformation of the BM is the "sharing" phase, where the network of the BM is consolidated into a	Aside from the "hurdles of transition" and the related stages of growth, Picken (2017): Table 1, p.589) declares no other signs of effective, proactive, and continuous communication seem relevant to cope with the transitions described by the CDM. Roome and Louche (2016) seem little interested in the idea of sharing information, until the last stage of the evolution of a new BM (p. 31). All of them may contradict these CDM claims, although both sets of	This underexplored field of shared learnings, improved communication, and assumed responsibilities of the "success adjudicator" is barely improved by these CDM assumptions. As they are originally stated, they relate to several traits and calls for action and recommendations that need further exploration Questions like what to share, with whom, communication tools, and how different alternatives improve/hinder learning or decision-making

(continued)

Table 1 (continued)

CDM rules	Evidence for	Evidence against	Remarks
	collaborative structure involving internal and external actors (Roome & Louche, 2016)	investigators say clearly that communication of a distinctive sense of direction is also essential for effective transitioning from idea to successful BM	are very much left unsolved and under-researched

functions within Ries' Lean Startup (LS) trademark interfacing two other different methodologies (Blank, 2013), totalling the three elements of LS:

- CusDev process. Within the largest context of LS, CusDev performs the core functions of LS (Ries, 2011: p17; Blank, 2013: p.6, Exhibit "listen to customers"). More specifically, it structures the process for "testing ... BMs' hypotheses—markets, customers, channels, pricing—and for turning those guesses into facts" (Blank & Dorf, 2012: p.22) with the aim of bringing out BMs that can repeat sales, and grow.
- Business model design. This is the methodology developed by Osterwalder and Pigneur (2010) to activate the BM construct as a nine-block template called the Business Model Canvas (BMC). BMC allows an easier way to depict how a business creates, delivers, and captures value. According to Blank and Dorf (2012), it "replaces static business plans" and serves as a route-map and dashboard when starting a business.
- Agile development (Beck, 2001; Blank, 2010; Dybå & Dingsøyr, 2008; Hui, 2013). Being a programming method, Agile was imported into the enhanced CusDev process by Blank (2007). It was later expanded by Ries (2011) (a programmer himself) with LS as this methodology brings concentration to LS. Whereas CusDev advocates for speed and urgency, Agile aims for "elimination of wasted time and resources by developing the product iteratively and incrementally" (Blank, 2013), using the stakeholders' feedback and knowledge gained through CusDev tasks. Agile covers those situations where the problem is known, but the solution is unknown. Blending it with CusDev, innovators can manage situations where the problem and the solution are unknown (Ries, 2009).

The CusDev process is conceptually detailed in many Blank and Dorf writings, mostly on their websites and other open-access sites. As they describe it, CusDev specifies the activities related to the development of a new business model (BM), blending activities related to the creation of a product with activities related to satisfying the needs of different stakeholders (Fig. 1). In terms of the value flow,

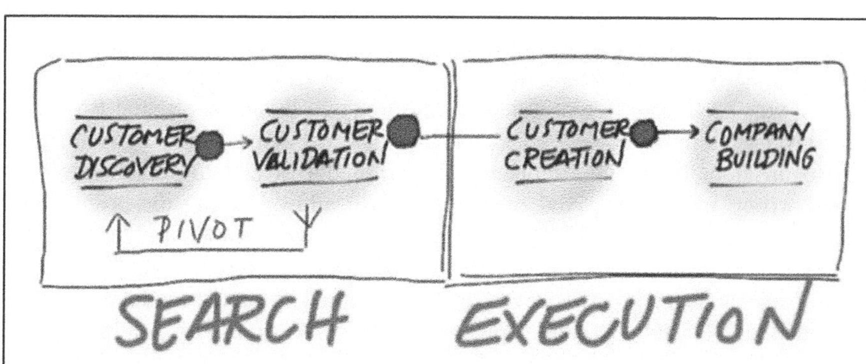

Fig. 1 The CusDev process. This is the original version from Blank (2015a, b)

CusDev presents an alternative (Blank, 2013) to structure value creation, value delivery, and value capture activities (Osterwalder & Pigneur, 2010) in situations where the problem and the solution are unknown. These activities have a common goal: the creation of a BM that repeats sales and grows. To achieve this goal, CusDev is built of four main sub-processes, or phases (Fig. 1).

The CusDev process in Fig. 1 shows two of its phases concentrating on searching for BMs that can sell repeatedly and grow, which is CusDev's ultimate goal. CusDev phases evolve through activity circles, or stages, that repeat cyclically until reaching a saturation point. With every turn of the cycle, Blank and Dorf posed one final question (pivot or proceed?), as if they explicitly would like to remark the importance of data validation and assessment in each cycle and their lining up toward the final goal of creating one (or more) BM that sells and grows. The last two sub-processes execute the BMs just created with thc aim of building a company.

(Refer to Data 1, Table A5 in Appendix 1 for the major constructs of CusDev sub-processes, and our elaboration on how the constructs affect the dimensions, or attributes, of selected models for conceptualizing sustainable business models (SBMs), eco-innovation, and the value cycle [Carrillo-Hermosilla et al., 2010; Blank & Dorf, 2012; Bocken et al., 2013; Boons & Lüdeke-Freund, 2013; Lüdeke-freund & Gold, 2016].)

One major insight from this CusDev, particularly when compared with conventional NPD, is that the CusDev is described as nonlinear. It is a sequence of circular phases with a point of return called the "pivot" (Ries, 2011: Chap. 8) after the Customer Validation stage. This return would happen if the uncovered BM cannot grow further (Fig. 1). Ries (2011) called this point *the pivot* to underscore the need for "changing the strategy" based on validated learning, when any of the BM sustaining hypotheses remains unproven, or is proven wrong, even when the other hypotheses may still be true, or not proven wrong. The CusDev design, away from a linear flow of actions or stages, may produce different BMs, in different layers, resulting in multilinear evolutions of the original BM as hypotheses become accepted/rejected (Fig. 2).

Figure 2 shows how innovators can choose their BM initial trajectory (grey arrow) extending from the center of the circle caping the cylinder. This circle represents a metaphor for the infinite goals that anyone starting a BM on day 0. The gray arrow clearly marks the desired trajectory from beginning to end goal. Then, pivots produce different trajectories. Some will leave the original BM in the same layer yet aiming for new goals. Other pivots will demand moving the original BM to different layers thus replacing the original BM due to valueholders and missed/wrong hypotheses. And other pivots will allow keeping the original BM and complement it with other BMs located in other layers of the cylinder.

The multidimensionality of BMs and their dynamism represented by Fig. 2 is much better grasped by CusDev than conventional business model innovation methods like New Product Development (NPD), which are linear in their assumptions. The layers of the cylinder in Fig. 2, and the trajectories within each layer, underscore the need for tools and practices that allow tracking of the complementarities and contradictions of the different BMs needed to satisfy stakeholders and

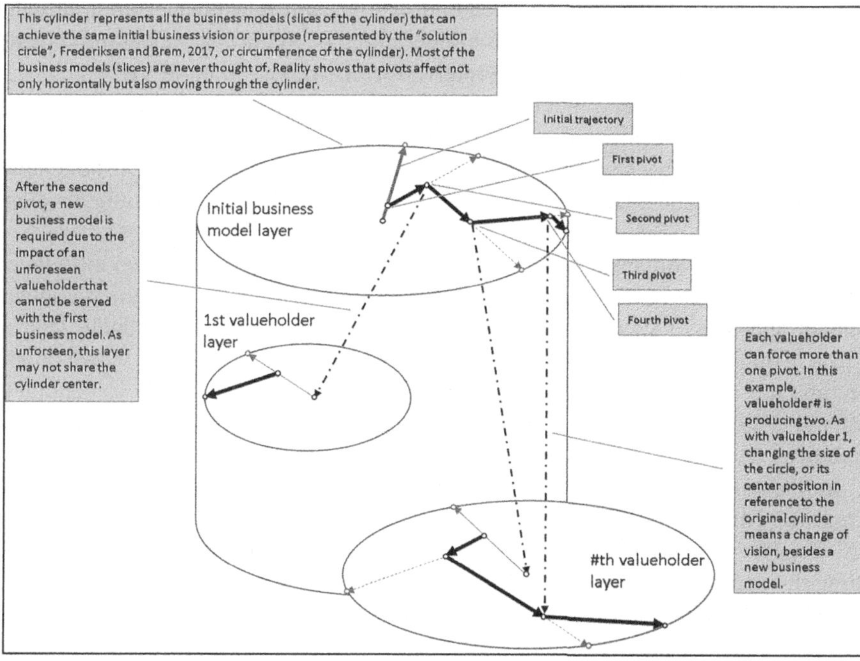

Fig. 2 Result of the application of the Customer Development process (Derived from Ries, 2011; Blank & Dorf, 2012; Frederiksen & Brem, 2017)

valueholders; many of these BMs are ephemeral, some reap immediate benefits, whereas others have long-term returns, and some run at a loss in economic terms simultaneously producing social and environmental benefits.

4 Discussion and Conclusions

This chapter purpose is threefold. One, we aim to identify the implications of the practice claims posed by the Customer Development Manifesto (CDM) for business models aiming for sustainability, or sustainable business models (SBMs). Second, we want to deepen on how SBMs might react to valueholders throughout their growth. Third, we ambition to understand how a practice like Customer Development (CusDev) encourages or nurtures sustainable business model innovation (SBMI). Having fulfilled our purpose, we have filled a gap in the extant literature, which overlooked CusDev capacity to function as way to innovate SBMs. CusDev (Blank, 2007; Blank & Dorf, 2012), as an integral part of the Lean Startup (LS) method (Ries, 2011), is frequently understood as just an alternative way to the conventional business model innovation. Blank and others grounded CusDev and LS as a faster, more focused methodology to build new businesses, following quasi-scientific validation of business underpinnings. With this paper, we contend that CusDev emerges as a sound alternative for SBMI, i.e., to build SBMs.

Following our chapter's first research question, we have connected the CDM with academic antecedents supporting the theories around business modelling and innovation. In addition, we have demonstrated the beneficial influence that a list of claims derived from practice, rigorously grounded, as the CDM can have on the development of SBMs.

Valueholders' influence and the reaction of SBMs to them have been introduced after the chapter's second question. The valueholder concept imagines relevant (salient and fringe) stakeholders (Mitchell et al., 1997) acting individually and at different times over the SBM growth. Beyond stakeholder groups, which might become too abstract (Hart & Sharma, 2004; Mattingly et al., 2004), valueholders emerge distinctively, compromising or supporting the SBMs triple-bottom-line viability.

In line with the chapter's third research question, the CDM introduced the use of CusDev. This practice is at the core of the search and grow strategies of the LS. CusDev has profoundly been deconstructed over the chapter exposing its strengths and weaknesses for SBMI. Using three case-study companies, we have provided actual examples of the application and implications of CusDev in real settings. These evaluations showcase how business models (BMs) built using CusDev balance (structural and cultural) dimensions (Stubbs & Cocklin, 2008), activities, and challenges (Geissdoerfer et al., 2016; Schaltegger et al., 2012) associated with a triple bottom line (Elkington, 2004; Joyce et al., 2015).

4.1 The Customer Development Process and Sustainability

In our opinion, CusDev appears soundly designed to manage the triple-layered BMC (Joyce & Paquin, 2016). Through each of its sub-processes, CusDev validates the integration of all elements and layers to build compatible BMs—this ability is actually being reported in several private and public sector contexts (Thompson, 2017). Additionally, CusDev consecutively addresses the needs of stakeholders and, even more importantly, of valueholders by developing adapted (sometimes temporary) BMs to serve them. As a result, CusDev produces multiple BMCs and BMs, which need to be managed simultaneously. This is achieved by using the CusDev dashboard—a list of metrics that builds on the BM validated hypotheses (Blank & Dorf, 2012).

From Fig. 2, anyone could have the wrong impression that pivots solve any situation posed by different valueholders, or other limitations of the new business. This is recognized by the recurring questions "pivot or persevere?" (Ries, 2011), or "pivot or proceed?" (Blank & Dorf, 2012), placed after each CusDev stage. Yet, these questions entail a deep, ongoing assessment process that includes three main areas to review after each CusDev stage:

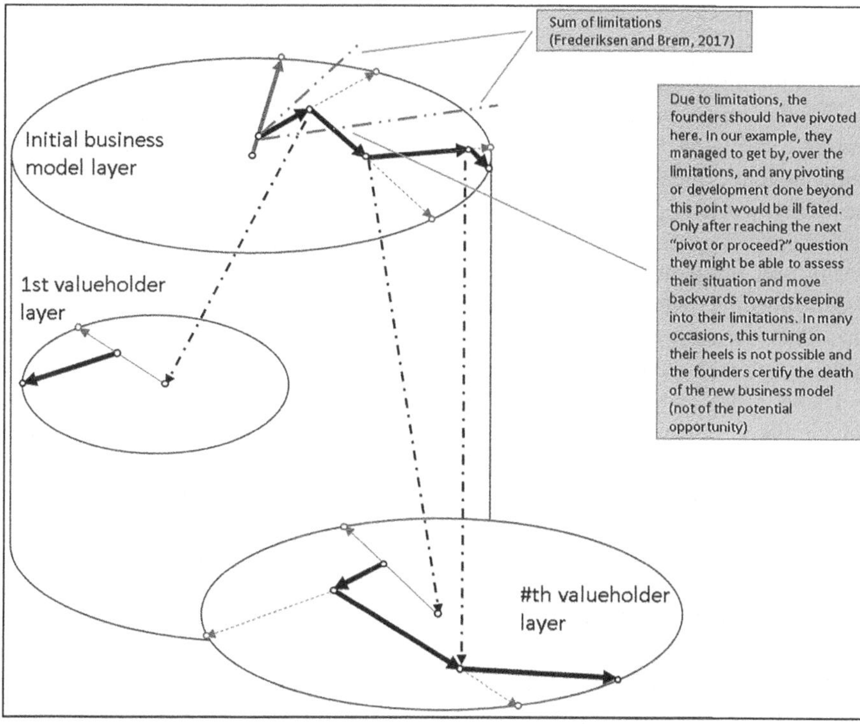

Fig. 3 Effect of limitations on a BM trajectory (Developed from Frederiksen & Brem, 2017)

- The founding team, and their abilities, motivation, resources, or commitments, to name just a few, akin to "who are you" (Sarasvathy, 2001).
- The validated learning through the CusDev phases, akin to "what do you know" (Sarasvathy, 2001).
- The friendly relations with the valueholders that can support the success of the BMs and the buying-in of the valueholders that can make the BM fail, akin to the "whom do you know" (Sarasvathy, 2001).

Any of the demands within the three assessment areas, if improperly addressed, can derail the search for, or the execution of, a new BM. Funding, time, premature scaling (trying to grow before it is time), and non-compliance are some examples of the limitations within the assessment areas and reasons behind BM failures beyond the remedy of the pivot. Figure 3 shows the effect of those limitations on the BM trajectories.

Using the metaphor in Fig. 2, the addition of limitations (dotted lines in Fig. 3) to the progression of BMs may result in their failure when their trajectory infringes any of those limitations. In Fig. 3, limitations emerge because of, e.g., limited funding and the need to comply with a CO2 regulation. These limitations fix the limits of the BM trajectories, becoming valueholders that need to be addressed with new BMs specifically designed to raise or eliminate the limitations. In Fig. 3, after the second pivot in the top layer, the BM trajectory would infringe the limit, which would imply the termination of the BM.

The theoretical implications of our research turn out as valuable. First, CusDev and its core claims listed in the CDM have been tested and successfully grounded using an ample and deep set of supporting and limiting academic literature. Second, the CDM becomes a rich source of requisites, or an outline that complements and makes explicit other influences currently investigated by behavioral researchers on design and implementation tools for SBMI. And third, the valueholder concept underscores the *dynamic* in the SBMs dynamic relationship with their triple bottom line. This dynamism can be further explored now regarding the emergence and influence of each valueholder through SBM growth.

Using CusDev-like methodologies allows startups to search for relevant valueholders, prioritize them, and accordingly solve the challenges driving their growth. Once valueholders are identified, CusDev guides the case startups through relentless learning cycles, where older challenges are replaced by new ones, and keeps the connection between planet, social, and economic bottom lines.

Organizing these challenges using Geissdoerfer and colleagues' activities and challenges (Geissdoerfer et al., 2016), the eco-innovation dimensions of Carrillo-Hermosilla et al. (2010) (see Data 1, Table A5, Appendix 1), the sustainable archetypes (N. Bocken et al., 2014) or the characteristics of sustainable BMs (Stubbs & Cocklin, 2008) helped us detect certain limitations of CusDev regarding its original design for SBMI. CusDev original structure lacks self-comparison data—ex ante/ex post comparison after running an innovation project. In its simplest form, CusDev might also induce confusion about leveraging different BMs under the same company. Lastly, CusDev original figures and exhibits weaken its potential to address its dynamic view of the SBM progress.

Being this a first approach to how CusDev can build SBMs more effectively, we believe that we have opened a relevant research field. We pose next steps toward the continuation of this promising field. We propose to confirm our preliminary validation of the CusDev methodology with cases, quantitative analyses, and even simulations. In addition, we encourage a more extensive description and explanation of SBMI challenges and activities from the CusDev perspective.

References

Anderson, C. (2006). *The long tail: Why the future of business is selling less of more*. Hyperion. https://doi.org/10.3359/oz0912041

Andreessen, M. (2007). *The only thing that matters*. THE PMARCA GUIDE TO STARTUPS. https://pmarchive.com/guide_to_startups_part4.html

Ansoff, H. I. (1957). Strategies for diversification. *Harvard Business Review, 35*, 113–124.

Aspara, J., Hietanen, J., & Tikkanen, H. (2010). Business model innovation vs replication: Financial performance implications of strategic emphases. *Journal of Strategic Marketing, 18*(1), 39–56. https://doi.org/10.1080/09652540903511290

Bajwa, S. S., Wang, X., Duc, A. N., & Abrahamsson, P. (2016). How do software startups pivot? Empirical results from a multiple case study. In A. Maglyas & A. L. Lamprecht (Eds.), *Software business, (ICSOB 2016)* (Vol. 240, pp. 169–176). https://doi.org/10.1007/978-3-319-40515-5_14

Bajwa, S. S., Wang, X., Nguyen Duc, A., & Abrahamsson, P. (2017). "Failures" to be celebrated: An analysis of major pivots of software startups. *Empirical Software Engineering, 22*(5), 2373–2408. https://doi.org/10.1007/s10664-016-9458-0

Baldassarre, B., Calabretta, G., Bocken, N. M. P., & Jaskiewicz, T. (2017). Bridging sustainable business model innovation and user-driven innovation: A process for sustainable value proposition design. *Journal of Cleaner Production, 147*, 175–186. https://doi.org/10.1016/j.jclepro.2017.01.081

Baumeister, H., Lichter, H., & Riebisch, M. (eds.). (2017). Agile processes in software engineering and extreme programming. In *18th International Conference, XP 2017 Cologne, Germany, May 22–26, 2017 Proceedings* (Vol. 283). Springer Open. https://doi.org/10.1007/978-3-319-57633-6.

Beck, K. (2001). *Manifesto for agile software development.* http://agilemanifesto.org/

Bengtsson, L., & Tavassoli, S. (2018). The role of business model innovation for product innovation performance. *Papers in Innovation Studies, Lund Unive* (2018/4). https://ideas.repec.org/p/hhs/lucirc/2018_004.html

Bertels, H. M., Koen, P. A., & Elsum, I. (2015). Business models outside the core: Lessons learned from success and failure. *Research-Technology Management, 58*(2), 20–29. https://doi.org/10.5437/08956308X5802294

Blank, S. (2007). *The four steps to the epiphany: Successful strategies for products that win* (3rd ed.). K&S Ranch.

Blank, S. (2010). *Why startups are agile and opportunistic—Pivoting the business model.* https://steveblank.com/2010/04/12/why-startups-are-agile-and-opportunistic-—pivoting-the-business-model/

Blank, S. (2012). *Nail the customer development manifesto to the wall.* SteveBlank.Com. https://steveblank.com/2012/03/29/nail-the-customer-development-manifesto/

Blank, S. (2013). Why the lean start-up changes everything. *Harvard Business Review, 91*, 64–72.

Blank, S. (2015a). *Innovation at 50x.* https://es.slideshare.net/sblank/innovation-at-50x-081515-51681670/54-Part_1Agile_EngineeringPart_2Part_3Elements

Blank, S. (2015b). *Organizational debt is like technical debt—but worse.* SteveBlank.Com. https://steveblank.com/2015/05/19/organizational-debt-is-like-technical-debt-but-worse/

Blank, S., & Dorf, B. (2012). *The startup owner's manual, the step-by-step guide for building a great company.* K&S Ranch. https://doi.org/10.1017/CBO9781107415324.004

Blank, S., & Newell, P. (2017). What your innovation process should look like. *Harvard Business Review, September 2017*, 1–2. https://hbr.org/2017/09/what-your-innovation-process-should-look-like

Block, Z., & MacMillan, I. C. (1985). Milestones for successful venture planning. *Harvard Business Review, 63*(5), 184–196. http://search.ebscohost.com.ezproxy.liv.ac.uk/login.aspx?direct=true&db=bth&AN=8500012358&site=ehost-live&scope=site

Bocken, N., Short, S. W., Rana, P., & Evans, S. (2014). A literature and practice review to develop sustainable business model archetypes. *Journal of Cleaner Production, 65*, 42–56. https://doi.org/10.1016/j.jclepro.2013.11.039

Bocken, N., Short, S., Rana, P., & Evans, S. (2013). A value mapping tool for sustainable business modelling. *Corporate Governance: The International Journal of Business in Society, 13*(5), 482–497. https://doi.org/10.1108/CG-06-2013-0078

Boons, F., & Lüdeke-Freund, F. (2013). Business models for sustainable innovation: State-of-the-art and steps towards a research agenda. *Journal of Cleaner Production, 45*, 9–19. https://doi.org/10.1016/j.jclepro.2012.07.007

Burke, A. E., Fitzroy, F. R., & Nolan, M. A. (2008). What makes a die-hard entrepreneur? Beyond the "employee or entrepreneur" dichotomy. *Small Business Economics, 31*(2), 93–115. https://doi.org/10.1007/s11187-007-9086-6

Burke, A., Fraser, S., & Greene, F. J. (2010). Multiple effects of business plans on new ventures. *Journal of Management Studies, 47*(3), 391–415.

Carrillo-Hermosilla, J., Del Río, P., & Könnölä, T. (2010). Diversity of eco-innovations: Reflections from selected case studies. *Journal of Cleaner Production, 18*(10–11), 1073–1083. https://doi.org/10.1016/j.jclepro.2010.02.014

Cassar, G., & Craig, J. (2009). An investigation of hindsight bias in nascent venture activity. *Journal of Business Venturing, 24*(2), 149–164. https://doi.org/10.1016/j.jbusvent.2008.02.003

Cavalcante, S., Kesting, P., & Ulhøi, J. (2011). Business model dynamics and innovation: (re)-establishing the missing linkages. *Management Decision, 49*(8), 1327–1342. https://doi.org/10.1108/00251741111163142

Chang, Y. C., Chang, H. T., Chi, H. R., Chen, M. H., & Deng, L. L. (2012). How do established firms improve radical innovation performance? The organizational capabilities view. *Technovation, 32*(7–8), 441–451. https://doi.org/10.1016/j.technovation.2012.03.001

Chawla, S. K., Hazeldine, M. F., Jackson, R. E., & Lawrence, R. J. (2007). Small business critical success factors and the legal form of the firm. *Journal of Business and Entrepreneurship, 19*(2).

Chesbrough, H. (2010). Business model innovation: Opportunities and barriers. *Long Range Planning, 43*, 354–363.

Christensen, C. M. (1997). *The innovator's dilemma: When new technologies cause great firms to fail*. HBS.

Christensen, C. M., Bartman, T., & van Bever, D. (2016). The hard truth about business model innovation. *Sloan Management Review, 58*(1), 31–40.

Cohen, W. M., & Levinthal, D. A. (1990). Absorptive capacity: A new perspective on learning and innovation. *Administrative Science Quarterly, 35*, 128–152. https://doi.org/10.2307/2393553

Crossan, M. M., & Apaydin, M. (2010). A multi-dimensional framework of organizational innovation: A systematic review of the literature. *Journal of Management Studies, 47*(6), 1154–1191. https://doi.org/10.1111/j.1467-6486.2009.00880.x

Cueto, J., & Zhang, S. (2014). *New venture decisions–A literature review of the entrepreneurial heuristics and biases*. 1–24. https://scholar.google.pl/scholar?hl=pl&q=New+Venture+Decisions+-+A+Literature+Review+of+the+Entrepreneurial+Heuristics+and+Biases&btnG=&lr=#0

Davila, A., Foster, G., Wilsz, J., Van Steele, J., Swanson, A. N., Armstrong, C., Chong, J., Ederhof, M., & Sarin, R. (2007). Management control systems in early-stage startup companies. *The Accounting Review, 82*(4), 907–937. https://doi.org/10.2308/accr.2007.82.4.907

Dennehy, D., Kasraian, L., Raghallaigh, P. O., & Conboy, K. (2016). Product market fit frameworks for lean product development. *R&D Management Conference 2016 "From Science to Society: Innovation and Value Creation," July*, 1–11.

Donaldson, T., & Preston, L. E. (1995). The stakeholder theory of the corporation: Concepts, evidence, and implications. *Academy of Management Review, 20*(1), 65–91.

Dorf, B. (2013). *When starting up, let the Customer Development Manifesto be your guide*. LinkedIn.Com. https://www.linkedin.com/pulse/20130405201046-7979133-when-starting-up-let-the-customer-development-manifesto-be-your-guide/

Drexler, A., Fischer, G., & Schoar, A. (2010). Keeping it simple : Financial literacy and rules of thumb keeping it simple: Financial literacy and rules of thumb. *CEPR Development Economics Workshop, 6*(October), 8–9. https://doi.org/10.1257/app.6.2.1

Duxbury, T. (2014). Improvising entrepreneurship. Technology innovation. *Management Review, 4*(7), 22–26.

Dybå, T., & Dingsøyr, T. (2008). Empirical studies of agile software development: A systematic review. *Information and Software Technology, 50*(9–10), 833–859. https://doi.org/10.1016/j.infsof.2008.01.006

Edison, H. (2017). *Lean internal startups: Empowering software product innovation in large companies* [Free University of Bozen-Bolzano, Italy]. https://doi.org/10.1016/j.jss.2017.09.034.

Edison, H., Smørsgård, N. M., Wang, X., & Abrahamsson, P. (2018). Lean internal Startups for software product innovation in large companies: Enablers and inhibitors. *Journal of Systems and Software, 135*, 69–87. https://doi.org/10.1016/j.jss.2017.09.034

Eisenmann, T., Ries, E., & Dillard, S. (2011). Hypothesis-driven entrepreneurship: The lean startup. *Harvard Business School Background Note 812–095, 44*(December), 1–23.

Elkington, J. (2004). Enter the triple bottom line. In A. Henriques & J. Richardson (Eds.), *The triple bottom line: Does it all add up* (p. 208). Routledge. https://doi.org/10.4324/9781849773348

Felin, T., & Foss, N. J. (2005). Strategic organization: A field in search of micro-foundations. *Strategic Organization, 3*(4), 441–455. https://doi.org/10.1177/1476127005055796

Forés, B., & Camisón, C. (2016). Does incremental and radical innovation performance depend on different types of knowledge accumulation capabilities and organizational size? *Journal of Business Research, 69*(2), 831–848. https://doi.org/10.1016/j.jbusres.2015.07.006

Frederiksen, D. L., & Brem, A. (2017). How do entrepreneurs think they create value? A scientific reflection of Eric Ries' lean Startup approach. *International Entrepreneurship and Management Journal, 13*(1), 169–189. https://doi.org/10.1007/s11365-016-0411-x

Freeman, R. E., Wicks, A. C., & Parmar, B. (2004). Stakeholder theory and "the corporate objective revisited.". *Organization Science, 15*(3), 364–369. https://doi.org/10.1287/orsc.1040.0066

Garrett, R. P., & Neubaum, D. O. (2013). Top management support and initial strategic assets: A dependency model for internal corporate venture performance. *Journal of Product Innovation Management, 30*(5), 896–915. https://doi.org/10.1111/jpim.12036

Garvin, D., Edmondson, A., & Gino, F. (2008). Is yours a learning organization? *Harvard Business Review, 86*(3), 109–116. 134. http://www.egatlearning.com/2011egrp3/reading/IsYoursaLearningOrganization.pdf

Gassmann, O., Frankenberger, K., & Csik, M. (2014). St. Gallen business model navigator. In Management of the fuzzy front end of innovation (18, 3). https://doi.org/10.1007/978-3-319-01056-4_7.

Geissdoerfer, M., Savaget, P., & Evans, S. (2016). The Cambridge business model innovation process. In G. Seliger, H. Kohl, & G. Oosthuizen (Eds.), 14th Global Conference on Sustainable Manufacturing GCSM 2016 (October, pp. 262–269). https://doi.org/10.1016/j.promfg.2017.02.033.

Giardino, C., Wang, X., & Abrahamsson, P. (2014). Why early-stage software Startups fail: A behavioral framework. In C. Lassenius & K. Smolander (Eds.), *Software business: Towards continuous value delivery* (Vol. 182, pp. 27–41).

Giesen, E., Berman, S. J., Bell, R., & Blitz, A. (2007). Three ways to successfully innovate your business model. *Strategy & Leadership, 35*(6), 27–33. https://doi.org/10.1108/10878570710833732

Giesen, E., Riddleberger, E., Christner, R., & Bell, R. (2010). When and how to innovate your business model. *Strategy & Leadership, 38*(4), 17–26. https://doi.org/10.1108/10878571011059700

Gino, F., Bohmer, R. M. J., Edmondson, A. C., Pisano, G. P., & Winslow, A. B. (2006). *Learning tradeoffs in organizations: measuring multiple dimensions of improvement to investigate learning-curve heterogeneity* (Issue June 2014).

Hambrick, D. C., & Mason, P. A. (1984). Upper echelons: The organization as a reflection of its top managers. *Academy of Management Review, 9*, 193–206. https://doi.org/10.2307/258434

Hamilton, R. H., & Hamilton, P. L. (2005). Timeless advice: Daniel Defoe and small business management. *Management Decision, 43*(10), 1304–1316. https://doi.org/10.1108/00251740510634886

Harms, R. (2015). Self-regulated learning, team learning and project performance in entrepreneurship education: Learning in a lean startup environment. *Technological Forecasting and Social Change, 100*, 21–28. https://doi.org/10.1016/j.techfore.2015.02.007

Harms, R., Kraus, S., & Reschke, C. H. (2007). Configurations of new ventures in entrepreneurship research: Contributions and research gaps. *Management Research News, 30*(9), 661–673. https://doi.org/10.1108/01409170710779971

Harms, R., Marinakis, Y., & Walsh, S. T. (2015). Lean startup for materials ventures and other science-based ventures: Under what conditions is it useful? *Translational Materials Research, 2*, 3. https://doi.org/10.1088/2053-1613/2/3/035001

Hart, S. L., & Sharma, S. (2004). Engaging fringe stakeholders for competitive imagination. *Academy of Management Executive*. https://doi.org/10.5465/AME.2004.12691227

Hui, A. (2013). Lean change: Enabling agile transformation through lean startup, Kanban, and Kotter an experience report. In *2013 AGILE CONFERENCE (AGILE)* (pp. 169–174). https://doi.org/10.1109/AGILE.2013.22.

Hurley-Hanson, A. E., Giannantonio, C. M., & Sudek, R. (2013). High-tech entrepreneurial careers: A model of stages, phases, and ages. *Journal of Business & Management, 19*(1), 11–24. http://search.ebscohost.com/login.aspx?direct=true&db=bth&AN=108357846&site=ehost-live

Im, K., & Cho, H. (2013). A systematic approach for developing a new business model using morphological analysis and integrated fuzzy approach. *Expert Systems with Applications, 40*(11), 4463–4477. https://doi.org/10.1016/j.eswa.2013.01.042

Jain, R. K. (2011). Entrepreneurial competencies: A meta-analysis and comprehensive conceptualization for future research. *Vision: The Journal of Business Perspective, 15*(2), 127–152. https://doi.org/10.1177/097226291101500205

Johnson, M. W., Christensen, C. M., & Kagermann, H. (2008). Reinventing your business model. *Harvard Business Review, December*, 1–10. https://doi.org/10.1111/j.0955-6419.2005.00347.x.

Joyce, A., & Paquin, R. L. (2016). The triple layered business model canvas: A tool to design more sustainable business models. *Journal of Cleaner Production, 135*, 1474–1486. https://doi.org/10.1016/j.jclepro.2016.06.067

Joyce, A., Paquin, R., & Pigneur, Y. (2015). The triple layered business model canvas A tool to design more sustainable business models. *ARTEM Organizational Creativity International Conference*, 33. https://blog.ssbmg.com/2015/04/17/the-triple-layered-business-model-canvas-a-tool-to-design-more-sustainable-business-models/.

Karadag, H. (2015). Financial management challenges in small and medium-sized enterprises: A strategic management approach. *EMAJ: Emerging Markets Journal, 5*(1), 26–40. https://doi.org/10.5195/EMAJ.2015.67

Khanagha, S., Volberda, H., & Oshri, I. (2014). Business model renewal and ambidexterity: Structural alteration and strategy formation process during transition to a cloud business model. *R&D Management, 44*(3), 322–340.

Kotler, P., Rackham, N., & Krishnaswamy, S. (2006). Ending the war between sales & marketing. *Harvard Business Review, 84*(7/8), 68–78. http://search.ebscohost.com/login.aspx?direct=true&db=buh&AN=21124747&site=ehost-live&scope=site

Lanning, M. J. (1998). *An introduction to the market-focused philosophy, framework and methodology called delivering profitable value* (Vol. 1, pp. 144–147). Perseus Publishing. http://kentblumberg.typepad.com/kent_blumberg/files/dpv_introduction.pdf

Leite, M., & Braz, V. (2016). Agile manufacturing practices for new product development: Industrial case studies. *Journal of Manufacturing Technology Management, 27*(4), 560–576. https://doi.org/10.1108/JMTM-09-2015-0073

Leonard-Barton, D. (1995). *Wellsprings of knowledge: Building and sustaining the sources of innovation*. Harvard Business School Press.

Linton, J., & Walsh, S. (2003). From bench to business. *Nature Materials, 2*(5), 287–289. https://doi.org/10.1038/nmat882

López, A., Contreras, R., & Espinosa, R. (2012). The impact of financial decisions and strategy on small business competitiveness. *Global Journal of Business Research, 6*(2), 93–104. http://papers.ssrn.com/sol3/papers.cfm?abstract_id=1945906

Lüdeke-freund, F., & Gold, S. (2016). Sustainable business model and supply chain conceptions towards an integrated perspective. In W. L. Bals & Tate (Eds.), *Implementing triple bottom line sustainability into global supply chains* (pp. 337–363). Greenleaf.

Makridakis, S., Hogarth, R. M., & Gaba, A. (2010). Why forecasts fail. What to do instead. *MIT Sloan Management Review, 51*(2), 83–90.

Mansoori, Y. (2017). Enacting the lean startup methodology. The role of vicarious and experiential learning processes. *International Journal of Entrepreneurial Behavior & Research, 23*(5), 812–838. https://doi.org/10.1108/IJEBR-06-2016-0195

March, J. G. (1979). The technology of foolishness. In *Decision Making in Ambiguity* (pp. 253–265).

Markides, C. (2006). Disruptive innovation: In need of better theory. *Journal of Product Innovation Management, 23*(1), 19–25. https://doi.org/10.1111/j.1540-5885.2005.00177.x

Markides, C. C. (1999). A dynamic view of strategy. *MIT Sloan Management Review, Spring*.

Mattingly, J. E., Post, J. E., Preston, L. E., & Sachs, S. (2004). Redefining the corporation: Stakeholder management and organizational wealth. *The Academy of Management Review*. https://doi.org/10.2307/20159063

Maurya, A. (2012). Running Lean: Iterate from Plan A to a Plan that works. In *O'Reilly Media*. https://doi.org/10.1126/sageke.2002.20.nw68.

Maurya, A. (2016). *Scaling lean: Mastering the key metrics for startup growth (issue October)*. Penguin Lcc Us.

McGrath, R. G. (2013). *The end of competitive advantage: How to keep your strategy moving as fast as your business*. Harvard Business Review.

McGrath, R. G. (2010). Business models: A discovery driven approach. *Long Range Planning, 43*(2–3), 247–261. https://doi.org/10.1016/j.lrp.2009.07.005

Merfeld, D. (2014). GE is in a startup state of mind. *Research Technology Management, 57*(6), 26–31. https://doi.org/10.5437/08956308X5706012

Miller, D. (1987). The genesis of configuration. *Academy of Management Review, 12*(4), 686–701.

Mintzberg, H. (1994). The fall and rise of strategic planning. *Harvard Business Review, January–February*, 48.

Mitchell, R. K., Wood, D. J., & Agle, B. (1997). Toward a theory of stakeholder identification and salience: Defining the principle of who and what really counts. *Academy of Management Review, 22*(4), 853–886. https://doi.org/10.5465/AMR.1997.9711022105

von Moltke, H. (1871). *Helmuth von Moltke the Elder*. Wikiquote. https://en.wikiquote.org/wiki/Helmuth_von_Moltke_the_Elder

Moogk, D. R. (2012). Minimum viable product and the importance of experimentation in technology startups. *Technology Innovation Management Review, 2*(3), 23–26. https://search.proquest.com/docview/1614473278?accountid=14475

Morris, M., Schindehutte, M., & Allen, J. (2005). The entrepreneur's business model: Toward a unified perspective. *Journal of Business Research, 58*(6), 726–735. https://doi.org/10.1016/j.jbusres.2003.11.001

Muller, A., Välikangas, L., & Merlyn, P. (2005). Metrics for innovation: Guidelines for developing a customized suite of innovation metrics. *IEEE Engineering Management Review, 33*(4), 66–72. https://doi.org/10.1109/EMR.2005.27017

Mullins, J. W., & Komisar, R. (2009). *Getting to plan B: Breaking through to a better business model*. Harvard Business School Press.

Nightingale, P., & Coad, A. (2014). Muppets and gazelles: Political and methodological biases in entrepreneurship research. *Industrial and Corporate Change, 23*(1), 113–143. https://doi.org/10.1093/icc/dtt057

Nonaka, I., Toyama, R., & Konno, N. (2000). SECI, Ba and leadership: A unified model of dynamic knowledge creation. *Long Range Planning, 33*(1), 5–34. https://doi.org/10.1016/S0024-6301(99)00115-6

O'Connor, G. C., Ravichandran, T., & Robeson, D. (2008). Risk management through learning: Management practices for radical innovation success. *Journal of High Technology Management Research, 19*(1), 70–82. https://doi.org/10.1016/j.hitech.2008.06.003

OECD. (2012). THE FUTURE OF ECO-INNOVATION: The role of business models in green transformation. In *OECD background paper, in Proceedings of the OECD/European Commission/Nordic Innovation Joint Workshop* (Issue January). https://doi.org/10.1037/0022-3514.85.4.768.

Osterwalder, A. (2004). The business model ontology—A proposition in a design science approach [UNIVERSITE DE LAUSANNE]. In *Business: Vol. Doctor*. https://doi.org/10.1111/j.1467-9310.2010.00605.x.

Osterwalder, A., & Pigneur, Y. (2010). Business model generation: A handbook for visionaries, game changers, and challengers. In *Wiley* (Vol. 30, Issue 5377). https://doi.org/10.1523/JNEUROSCI.0307-10.2010.

Patz, M. (2013). *Lean startup – Adding an experimental learning perspective to the entrepreneurial process*. Universiteit Twente, Technische Universität Berlin.

Pease, J. F. (2014). The lean design for the developing world method: a novel lean market-based product design methodology for developing world markets that benefits consumers and companies. *Proceedings of the ASME International Design Engineering Technical Conferences and Computers And Information In Engineering Conference*, VOL 2A.

Peralta, A., Carrillo-Hermosilla, J., & Crecente, F. (2018). *Alternatives for the eco-innovation of business models: A conceptual reference to valueholders* (2/2018; SERIE DOCUMENTOS DE TRABAJO). https://www.researchgate.net/publication/325336053_Alternatives_for_the_eco-innovation_of_business_models_a_conceptual_reference_to_valueholders

Phelan, S. E. (2015). The complexity of opportunity. *Academy of Entrepreneurship Journal, 21*(2), 55–72.

Picken, J. C. (2017). From startup to scalable enterprise: Laying the foundation. *Business Horizons, 60*(5), 587–595. https://doi.org/10.1016/j.bushor.2017.05.002

Powell, J. D., & Bitner, L. N. (1992). Specifying success measures within a performance domain. *Journal of Business and Entrepreneurship, 4*(3).

Rachleff, A. (2017). *How to find product market fit*. https://drt.fm/andy-rachleff

Rackham, N. (1999). *Rethinking the sales force: Redefining selling to create and capture customer value*. McGraw-Hill Education Ltd. http://www.altfeldinc.com/pdfs/salesforce.pdf

Rasmussen, E. S., & Tanev, S. (2015). The emergence of the lean global Startup as a new type of firm. *Technology Innovation Management Review*, 12–19.

Rasmussen, E. S., & Tanev, S. (2016). Lean start-up: Making the start-up more successful. In F. Pacheco-Torgal, E. Rasmussen, C.-G. Granqvist, V. Ivanov, A. Kaklauskas, & S. Makonin (Eds.), *Start-up creation: The smart eco-efficient built environment* (pp. 39–56). Elsevier Ltd.. https://doi.org/10.1016/B978-0-08-100546-0.00003-0

Rieple, A., & Pisano, P. (2015). Business models in a new digital culture: The open long tail model. *Symphonya, 2*, 75–88. https://doi.org/10.4468/2015.2.06pisano.rieple

Ries, E. (2009). *Customer development engineering*. Www.Startuplessonslearned.com. http://www.startuplessonslearned.com/search?q=customer+development

Ries, E. (2011). *The lean startup: How today's entrepreneurs use continuous innovation to create radically successful businesses* (1st ed.). Crown Business.

Rissanen, T., & Sainio, L.-M. (2016). Business model experimentation in incumbent and startup companies. *XXVII ISPIM Innovation Conference, June*, 1–12.

Rompho, N. (2018). Operational performance measures for startups. *Measuring Business Excellence, 22*, 31–41. https://doi.org/10.1108/MBE-06-2017-0028

Roome, N., & Louche, C. (2016). Journeying toward business models for sustainability: A conceptual model found inside the black box of organisational transformation. *Organization and Environment, 29*, 1. https://doi.org/10.1177/1086026615595084

Saaty, T. L. (2008). Decision making with the analytic hierarchy process. *International Journal of Services Sciences, 1*(1), 83. https://doi.org/10.1504/IJSSCI.2008.017590

Sai Manohar, S., & Pandit, S. R. (2013). Core values and beliefs: A study of leading innovative organizations. *Journal of Business Ethics, 125*(4), 667–680. https://doi.org/10.1007/s10551-013-1926-5

Sarasvathy, S. D. (2001). Causation and effectuation: Toward a theoretical shift from economic inevitability to entrepreneurial contingency. *Academy of Management Review, 26*(2), 243–263. https://doi.org/10.5465/AMR.2001.4378020

Schaltegger, S., Lüdeke-Freund, F., & Hansen, E. G. (2012). Business cases for sustainability: The role of business model innovation for corporate sustainability. *International Journal of Innovation and Sustainable Development, 6*(2), 95–119. https://doi.org/10.1504/IJISD.2012.046944

Sebastian Quintero. (2017). *Dissecting startup failure rates by stage*. TowardsDataScience.Com. https://towardsdatascience.com/dissecting-startup-failure-by-stage-34bb70354a36

Serrador, P., & Pinto, J. K. (2015). Does agile work?—A quantitative analysis of agile project success. *International Journal of Project Management, 33*(5), 1040–1051. https://doi.org/10.1016/j.ijproman.2015.01.006

Sharma, B. (2004). Marketing strategy, contextual factors and business performance: An investigation of their relationship. *Marketing Intelligence & Planning*.

Shepherd, D. A., Haynie, J. M., & McMullen, J. S. (2012). Confirmatory search as a useful heuristic? Testing the veracity of entrepreneurial conjectures. *Journal of Business Venturing, 27*(6), 637–651. https://doi.org/10.1016/j.jbusvent.2011.06.002

Sitkin, S. B. (1992). Learning through failure: The strategy of small losses. In B. M. Staw & L. L. Cummings (Eds.), *Research in organizational behavior* (pp. 231–266). JAI Press.

Spear, S., & Bowen, H. K. (1999). Decoding the DNA of the Toyota Production System. *Harvard Business Review, 77*(5), 96–106. https://doi.org/http://search.ebscohost.com/login.aspx?direct=true&db=buh&AN=2216294&site=ehost-live

Stalnaker, C. K. (2002). Making the transition from startup to normal operations. *Professional Safety, 47*(11).

Stampfl, G., Prügl, R., & Osterloh, V. (2013). An explorative model of business model scalability. *International Journal of Product Development, 18*(3/4), 226. https://doi.org/10.1504/IJPD.2013.055014

Stubbs, W., & Cocklin, C. (2008). Conceptualizing a "sustainability business model.". *Organization & Environment, 21*, 103–127. https://doi.org/10.1177/1086026608318042

Sutton, S. M. (2000). The role of process in software start-up. *IEEE Software, 17*(4), 33–39. https://doi.org/10.1109/52.854066

Tanev, S. (2017). Is there a lean future for global startups ? *Technology Innovation Management Review, 7*(5), 6–15.

Täuscher, K., & Abdelkafi, N. (2016). Modelling the Lean Startup: A simulation tool for entrepreneurial growth decisions. *Proceedings of the 16th EURAM Conference, June*.

Teece, D. J. (2010). Business models, business strategy and innovation. *Long Range Planning, 43*(2–3), 172–194. https://doi.org/10.1016/j.lrp.2009.07.003

Teece, D., & Pisano, G. (1994). The dynamic capabilities of firms: An introduction. *Industrial and Corporate Change, 3*(3), 537–556. https://doi.org/10.1093/icc/3.3.537-a

Thompson, L. (2017). *Georgetown University students get military buy-in on novel approach to solving security problems*. Forbes.Com. https://www.forbes.com/sites/lorenthompson/2017/03/21/georgetown-university-students-get-military-buy-in-on-novel-approach-to-solving-security-problems/#2bc283246afc

Tierney, R., Groen, A. J., Harms, R., Luizink, M., Hetherington, D., Stewart, H., Walsh, S. T., & Linton, J. (2012). Managing highly flexible facilities: An essential complementary asset at risk. *International Journal of Entrepreneurial Behavior & Research, 18*(2), 233–255. https://doi.org/10.1108/13552551211204238

Tornikoski, E., & Renko, M. (2014). Timely creation of new organizations: The imprinting effects of entrepreneurs' initial founding decisions. *Management Aims, 17*(3), 193–213.

Treacy, M., & Wiersema, F. (1993). Customer intimacy and other value disciplines customer intimacy and other value disciplines. *Harvard Business Review, 71*(9301), 84–93. https://doi.org/10.1225/93107

Trimi, S., & Berbegal-Mirabent, J. (2012). Business model innovation in entrepreneurship. *International Entrepreneurship and Management Journal, 8*(4), 449–465. https://doi.org/10.1007/s11365-012-0234-3

Tripp, J. F., Saltz, J., & Turk, D. (2018). Thoughts on current and future research on agile and lean: Ensuring relevance and rigor. *Proceedings of the 51st Hawaii International Conference on System Sciences, 9*, 5465–5472.

Unterkalmsteiner, M., Abrahamsson, P., Nguyen-duc, A., Baltes, G. H., Conboy, K., Dennehy, D., Sweetman, R., Edison, H., Shahid, S., Wang, X., Garbajosa, J., Gorschek, T., Hokkanen, L., Lunesu, I., Marchesi, M., Morgan, L., Selig, C., Oivo, M., Shah, S., & Kon, F. (2016). Software Startups—A research agenda. *E-Informatica Software Engineering Journal, 10*(1), 1–28. https://doi.org/10.5277/e-Inf160105

Verhoeven, B., & Johnson, L. W. (2017). Business model innovation portfolio strategy for growth under product-market configurations. *Journal of Business Models, 5*(1), 35–50. http://search.ebscohost.com/login.aspx?direct=true&db=bth&AN=126309379&site=ehost-live

von Hippel, E. (2013). *Understanding lead user research principles*. Lead User Project Handbook. https://evhippel.files.wordpress.com

Von Hippel, E. (1986). Lead users: An important source of novel product concepts. *Management Science, 32*(7), 791–805. https://doi.org/10.1017/CBO9781107415324.004

de Vries, M. F. R. K. (1996). The anatomy of the entrepreneur: Clinical observations. *Human Relations, 49*(7), 853–883.

Weissbrod, I., & Bocken, N. M. P. (2016). Developing sustainable business experimentation capability—A case study. *Journal of Cleaner Production*, 1–14. https://doi.org/10.1016/j.jclepro.2016.11.009

Wirtz, B. W., Pistoia, A., Ullrich, S., & Göttel, V. (2016). Business models: Origin, development and future research perspectives. *Long Range Planning, 49*(1), 36–54. https://doi.org/10.1016/j.lrp.2015.04.001

Yang, M., Evans, S., Vladimirova, D., & Rana, P. (2017). Value uncaptured perspective for sustainable business model innovation. *Journal of Cleaner Production, 140*, 1794–1804. https://doi.org/10.1016/j.jclepro.2016.07.102

Yang, X., Sun, S. L., & Zhao, X. (2018). Search and execution: Examining the entrepreneurial cognitions behind the lean startup model. *Small Business Economics*, 1–13. https://doi.org/10.1007/s11187-017-9978-z

Ye, Q. (2016). Effectual approaches and entrepreneurship outcome: From a perspective of behavioral biases. *Journal of Small Business & Entrepreneurship, 28*(5), 401–411. https://doi.org/10.1080/08276331.2016.1209028

Yin, R. K. (2009). Case study research : Design and methods. In L. Bickman & D. J. Rog (Eds.), *Applied social research methods series.*; Vol. Applied So. SAGE Publications. https://doi.org/10.1097/FCH.0b013e31822dda9e

York, J. L., & Danes, J. E. (2014). Customer development, innovation, and decision-making biases in the lean startup. *Journal of Small Business Management, 24*(2), 21–39. https://doi.org/10.1111/j.1540-627X.2011.00322.x

Yuan, X., Guo, Z., & Fang, E. (2014). An examination of how and when the top management team matters for firm innovativeness: The effects of TMT functional backgrounds. *Innovation: Management, Policy and Practice, 16*(3), 323–342. https://doi.org/10.1080/14479338.2014.11081991

Zott, C., Amit, R., & Massa, L. (2011). The business model: Recent developments and future research. *Journal of Management, 37*(4), 1019–1042. https://doi.org/10.1177/0149206311406265

Part II
Southwest European Perspective

The Implementation of Emerging Clean Technologies and Circular Value Chains: Challenges from Three Cases of By-Product Valorization

Laura Carraresi and Stefanie Bröring

Abstract Clean technologies enable—through cascading processing of agricultural waste and by-products—unlocking the full potential of biomass and pursuing value-added pathways beyond energy, i.e., bio-based fine chemicals, food ingredients, and pharmaceuticals. Likewise, alternative sources of biomass generate a range of fuels and building blocks for the chemical industry. These new industrial solutions result from boundary spanning applications of clean technologies across hitherto separated sectors triggering the emergence of new value chains. This chapter reviews three cases of emerging clean technologies: (a) secondary metabolites' extraction from tomato by-products, (b) phosphate recovery from rapeseed oil press-cakes, and (c) alternative sources of biomass (lignocellulose) for a new biorefinery concept. We aim to outline the challenges associated with the implementation of these technological innovations. An exploratory case-study method was applied to enable cross-case comparison. Data were derived from expert interviews with value-chain actors. The results revealed several barriers to the uptake of emerging clean technologies. Interviewees showed a widespread hesitation to redesign their current business models to implement novel clean technologies; this hesitation was associated mainly with uncertainty about economic viability and the long time span to the breakeven point. Regulatory issues and concerns regarding process efficiency and product quality were also highlighted.

Supplementary Information The online version contains supplementary material available at [https://doi.org/10.1007/978-3-031-08313-6_5].

L. Carraresi (✉)
Department of Innovation, Sensory and Consumer Science, Division Food, Nofima AS, Ås, Norway
e-mail: laura.carraresi@nofima.no

S. Bröring
Center for Entrepreneurship, Innovation and Transformation (CEIT), Chair for Entrepreneurship and Innovative Business Models, Ruhr-University Bochum, Bochum, Germany
e-mail: stefanie.broering@ruhr-university-bochum.de

V. Prokop et al. (eds.), *Business Models for the Circular Economy*, Sustainability and Innovation, https://doi.org/10.1007/978-3-031-08313-6_5

1 Introduction

The global challenge of climate change and the scarcity of fossil resources (Schmid et al., 2012) are currently creating, on the one side, an increasing need to reduce greenhouse gas emissions and, on the other side, the need to replace non-renewable resources with alternative and sustainable ones. These trends translate into a shift from a fossil-based, linear economy to a circular economy (Golembiewski et al., 2015; McCormick & Kautto, 2013), reflecting an overall system change toward sustainability (Dinesh et al., 2021; Patterson et al., 2017). This transformation increasingly requires companies to develop "system innovations." Such innovations are focused on "radically reconfiguring systems" (Dinesh et al., 2021, p. 2) and are disruptive rather than incremental. The systemic implementation of new clean technologies is necessary and works to catalyze the transformation of current practices in agriculture and food systems (Dinesh et al., 2021; Augustin et al., 2021; El Bilali, 2019; Leeuwis et al., 2021). Agriculture plays a growing innovative role in this scenario by providing new bio-based raw materials for industry, materials that are leveraged through the cascading use of biomass enabled by emerging technologies (Rotolo et al., 2015; Müller & Laibach, 2022; OECD, 2014). An emerging technology is a radical and fast-expanding technology that has relevant consequences on the configuration of the value chain, the synergies among actors and institutions, and the knowledge generation process. Since its biggest impact will only materialize in the future, in its emergence phase, it is still characterized by uncertainty and ambiguity (Rotolo et al., 2015).

In the bioeconomy context, some emerging technologies allow the extraction of fine chemicals from agri-food waste and by-products, and these chemicals have valuable applications as food additives, fertilizers, cosmetics, and pharmaceuticals (Müller & Laibach, 2022; Wesseler & Zilberman, 2021; Kardung et al., 2021; Cherubini, 2010). Likewise, alternative sources of biomass (e.g., perennial plants not competing with food production) can generate a range of fuels, materials, and building blocks for the chemical industry (Parisi, 2018; Clomburg et al., 2017; Kardung et al., 2021; FitzPatrick et al., 2010). Being less damaging on the environment than conventional technologies, they can be considered "clean" (Clift, 1995). Although the implementation of these emerging clean technologies is highly necessary to meet global sustainability development goals (SDGs), it remains challenging, due mainly to the lack of connections across distant sectors of the economy. Increasing ties among firms, additional value chain loops to manage biomass flows, and cross-industry networks are needed to restructure value chains toward increasing circularity (Bröring et al., 2020, Donner et al., 2020; Sheppard et al., 2011; Carraresi et al., 2018; Bröring & Cloutier, 2008). Additionally, the implementation of these technologies brings a lack of know-how in companies operating in distant sectors (Bierly & Chakrabarti, 2001), high-risk perception (Ekman et al., 2013; Stern & Valero, 2021), high switching costs (Aghion et al., 2019; Geroski, 2000), the absence of industry standards, long regulatory approval processes (Scarlat et al., 2015), the absence of downstream processing technologies (Mohan et al., 2016), and an overall high degree of uncertainty (Rotolo et al., 2015).

Although the issues facing emerging bio-based technologies are widespread, such challenges have been discussed in the literature mostly at a macro level within the debate on the overall food system's shift toward sustainability (Conti et al., 2021; Dinesh et al., 2021; Leeuwis et al., 2021). Literature exploring barriers and obstacles to technology implementation at the micro and meso levels is still scant (Carraresi et al., 2018). This gap is worth investigating, as systemic change can happen only if its components, namely, companies (micro level) and value chains (meso level), are also seeking change. Within the scope of this gap in the literature, emerging technologies aimed at valorizing agricultural by-products and alternative biomass sources are already available, and they constitute a basis on which to build a solid staircase toward sustainable processes (Chemat et al., 2020; Müller & Laibach, 2022).

Against this background, the present chapter aims to review three cases of emerging bio-based technologies, outline the associated challenges, highlight what is needed to foster their implementation, and provide managerial and policy recommendations.

The three emerging technologies analyzed in this chapter enable different uses of agricultural by-products and alternative biomass. The cases can be described as follows: (a) a process to stress tomato plants in greenhouses and extract secondary metabolites contained in the waste, which exhibit high chemical diversity and several biological activities valuable for human health (Chua, 2013); (b) a process of the release and recovery of phosphorus—a scarce and non-renewable resource—from rapeseed oil press cakes to produce bio-based polyphosphate before supplying press cakes to the feeding industry; and (c) a biorefinery process that extracts value-added compounds (e.g., lignin, hemicellulose, and different sugars) from perennial plants cultivated on marginal soils for use as a biomass source.

This chapter seeks to shed light on the challenges arising at both the value chain and firm levels when a novel clean technology is still in its early phase of development. The chapter thus contributes to theory development of the research body on clean technologies and renders managerial insights to support both researchers and managers in foreseeing implementation challenges, as well as possible market bottlenecks and drivers, enabling them to target and adapt the technology accordingly. The chapter also contributes to raising firms' awareness of the availability of novel bio-based technologies and valorization opportunities and provides managers and entrepreneurs with a blueprint for decision-making for future applications of emerging clean technologies.

In the following section, a review of the extant literature provides an overview of the potential challenges that companies might encounter when engaging with new clean technologies emerging at the interface of adjacent sectors. In Sect. 3, the methodology is presented, introducing the case studies, the related projects, and the interviews that have been carried out. In Sect. 4, the results of the interviews are reported, outlining the challenges that have been mentioned by respondents. In Sect. 5, the results are discussed in light of extant literature on the topic, and some managerial recommendations are provided. The conclusions are summarized in Sect. 6.

2 Challenges for the Implementation of Emerging Technologies in the Bioeconomy

Several challenges relative to the potential implementation of emerging technologies and the consequent restructuring of value chains have been underlined in the literature so far (Table S1). These challenges can be grouped into five categories: organizational, regulatory, economical, geographical, and product-quality related.

Organizational challenges relate to the necessity of developing appropriate competencies, acquiring missing knowledge, and developing new skills to implement innovations that share technology platforms across different sectors (Sheppard et al., 2011; Bai et al., 2017). Meeting these needs is challenging for three reasons which are to some degree also reinforcing each other: first, firms are often insufficiently able to understand and acquire knowledge from cognitively distant industries (Nooteboom, 2000) because they have a limited absorptive capacity (Cohen & Levinthal, 1990) and are caught in their knowledge corridors (Marvel et al., 2020); second, they tend to be anchored to their current business model (Laudien & Daxbock, 2016; Sydow et al., 2009); and third, they often resist to learn about and invest in radically new technologies, ending up stuck in their current technological setup, or a so-called technological lock-ins (Newton et al., 2020). The first factor limits firms' access to appropriate networks and, in turn, firms' capacity to recognize opportunities and adopt innovations that might arise in other sectors, which are only apparently distant (Aid et al., 2017; Golembiewski et al., 2015). In short, limited awareness and knowledge of technological advances can reduce a firm's capability to recognize market opportunities and identify good potential partners. The second factor increases the innovation implementation gap, as firms interested in adopting new technologies in the bioeconomy domain would often need to change or reinvent their business model, which is not always feasible (Kirchherr et al., 2018; Golembiewski et al., 2015; Johnson et al., 2008; Carraresi & Bröring, 2021). The third factor makes it difficult for firms to invest in new equipment and adapt their extant practices, hampering the development of alternative clean technologies (Conti et al., 2021).

Regulatory challenges concern a lack of industry standards. New bio-based technologies might use or produce materials with different properties from the conventional ones that must comply with not-yet-in-place regulations before entering the market (Golembiewski et al., 2015). In other cases, regulation is unclear to companies, unstable, or misaligned across different industrial or geographical markets (Peters & Buijs, 2022; Dangelico & Pontrandolfo, 2015; Conti et al., 2021).

Economic challenges relate to the efficiency of implementing emerging technologies. Despite their many benefits regarding the implementation of more sustainable pathways, clean technologies developed to valorize agricultural by-products are often not efficient enough and thus lack economic viability due to yet lower production costs of the current fossil-based system. Therefore, achieving efficiency and profitability with such technologies remains the key issue (Kirchherr et al., 2018; Maity, 2015; Maaß et al., 2014; Sheppard et al., 2011). These new clean technologies are still not fully competitive, and although bio-based products in some high-value applications can claim a premium price, they are still far from enabling economies of scale (Simon et al., 2016). In the specific case of using alternative biomass sources for a novel lignocellulosic biorefinery, the long establishment periods of the perennial plants used as sources of feedstock (e.g., up to 4 years in the case of Miscanthus) are a big challenge (Caslin et al., 2010). Thus, farmers would need to bear high opportunity costs in the first years of implementing this technology. The market for bio-based ingredients, materials, and products is still ambiguous and customer demand for them still lacking (Peters & Buijs, 2022; Alblas et al., 2014; Driessen et al., 2013; Kirchherr et al., 2018).

Geographical challenges are associated with logistical changes in the value chain. Suppliers of agricultural by-products or alternative biomasses are often located far away from each other and from processors, requiring chain actors to modify their logistics, centralize the supply, and integrate backward or forward (Donner et al., 2020).

Finally, challenges arise in achieving standardized *quality in the final product* (e.g., novel ISO norms are needed for bio-based products). Because by-products are often collected from several suppliers, they can have different properties, which then affect the features of the final product (Donner et al., 2020; Simon et al., 2016). In the case of alternative biomass sources, biomass has a high degree of functionality, so different biomass sources require unique reaction conditions (FitzPatrick et al., 2010), leading to feedstock of different quality and, consequently, to biorefinery products (e.g., lignin) with heterogeneous characteristics. Therefore, downstream chain actors might need to tackle these challenges, for example, by investing in further refinement processes, when they seek to deliver lignin-derived chemicals with high purity (Beckham et al., 2016).

3 Methodology

Although the number of studies in the literature investigating the emergence of new value chains in the bioeconomy domain is still limited, examples of emerging technologies are available and could be further developed and adapted for several industrial applications. This chapter seeks to review three cases of novel clean technologies developed in Germany, in the region of North-Rhine-Westphalia, and promoted by interdisciplinary research projects (funded by the Bioeconomy Science

Center – BioSC[1]) we were contributing to, building on the work of Carraresi et al. (2018) and Junker-Frohn et al. (2019). These cases have been analyzed following the exploratory case-study method (Eisenhardt & Graebner, 2007; Glaser & Strauss, 1967; Yin, 1994) and are hereby compared and discussed. The cases are selected by means of theoretical sampling in terms of value chain design, challenges arising from the emergence of new organizational structures, willingness to adopt new technologies, and opportunities offered by the valorization of raw materials. The aim was to develop an overarching framework of analysis for the phenomenon of novel clean technologies and their application in emerging circular value chains.

3.1 Description of the Cases

3.1.1 Case A: Extraction of Secondary Metabolites from Tomato By-Products

The first case study originates from the BioSC research project "Induction of secondary metabolites in tomato by-products for extraction and economic evaluation of the model process" (InducTomE), the aim of which is to identify abiotic stress treatments that induce secondary metabolites in green tomato by-products for further extraction and commercialization. The project targets two secondary metabolites, rutin and solanesol, which are health-promoting active components that can be used as ingredients in products such as herbal medicines, dietary supplements, and cosmetics (Pleissner et al., 2016; Kirby & Keasling, 2009; Yan et al., 2015).

The valorization of horticultural by-products could reduce the production of oil-based chemicals and open new market opportunities for farmers and companies (Wensing et al., 2019; Boehlje & Broering, 2011). Farmers might provide their biomass residuals to biorefineries or industrial partners and, in turn, diversify their income sources, whereas industrial companies could develop innovative products based on sustainably produced ingredients and, thus, charge a price premium for differentiation.

However, adopting this technological innovation to obtain these compounds from tomatoes requires adjustments within the conventional agri-food value chain (Fig. 1). This innovation requires tomato growers to perform an abiotic stress treatment in the greenhouse to induce stronger secondary metabolite production (e.g., by drought or nutrient deficiency). Subsequently, green biomass residuals must be harvested, dried, and delivered to chemical companies for extraction of the target components, rutin and solanesol. The ingredients can then be provided to

[1]The North Rhine-Westphalia strategy project BioSC has been launched in 2013 in the framework of the whole bioeconomy strategy of the German state North Rhine-Westphalia. It is a prototype model for a regional and transdisciplinary funding system which facilitates the cooperation among research institutions in the region under the common aim of fostering the bioeconomy. For further details, please visit the website: www.biosc.de

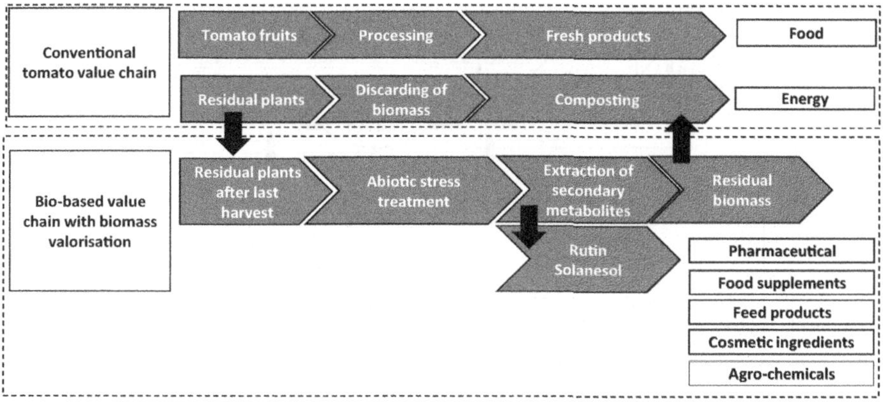

Fig. 1 Bio-based tomato value chain with biomass valorization. Source: own adaptation from Junker-Frohn et al. (2019)

downstream processors for the production of consumer goods, such as cosmetics or dietary supplements. Overall, such changes are likely to create challenges for different value chain agents associated with the implementation of novel processes (Scarlat et al., 2015). Therefore, farmers, chemical companies, and processors of consumer goods need to be willing to adopt a proposed process to generate a new bio-based value chain.

3.1.2 Case B: Phosphorus Release and Recovery from Rapeseed Oil Press Cakes

This case study emerges from the BioSC research project "Efficient phosphate recovery from agro-waste streams by enzyme, strain, and process engineering" (P-ENG), the aim of which is to obtain phosphorus through the hydrolysis of plant phytate in rapeseed oil press cakes. This would avoid the need for phosphorus extraction from rock mines for polyphosphate synthesis and decrease the phosphate content of agricultural waste streams. In this context, the current phosphate value chain would encounter organizational changes, so companies potentially interested in this technology need to overcome several challenges in the implementation process (Carraresi et al., 2018).

This new clean technology enables the mobilization and recovery of phosphate contained in rapeseed oil press cakes through the addition of an improved variant of the phytase enzyme, which can convert phosphate to polyphosphate of industrial value. In this way, rapeseed oil press cakes could be used not only as feed but also as a phosphorus source, making them more valuable in two ways: through the lower content of phosphate in the feed (and consequent better digestibility for monogastric animals) and through additional industrial applications (i.e., polyphosphate for fertilizers, food additives, and detergents). Other advantages of this technology are

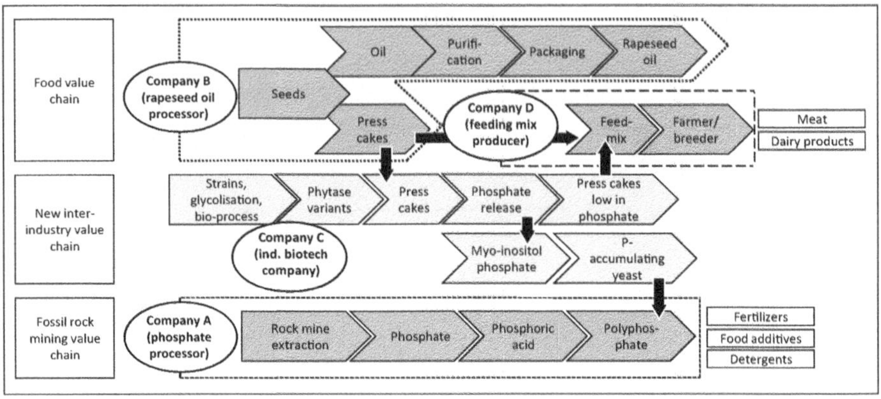

Fig. 2 Bio-based polyphosphate value chain. Source: Carraresi et al. (2018)

related to the reduction of water eutrophication and over-fertilization due to agricultural waste streams (Shilton & Blank, 2012), the higher purity level of the phosphorus produced (no residues of heavy metals), the elimination of the need to supplement trace elements in the feeding mix (as they are released alongside phosphorus), and the bio-based origin of the obtained polyphosphate.

The implementation of this process requires that different value chains (oil press cakes, enzymes, polyphosphates), which are currently separate, establish new boundary spanning business-to-business relationships, generating a potential new value chain (Fig. 2). For instance, instead of being delivered straight to the feeding industry as in the conventional process, the rapeseed oil press cakes would go through an extra loop, being treated with a variant of phytase-releasing phosphate to obtain polyphosphate, which would be used in turn for all targeted applications. Thus, the current rapeseed oil value chain would become connected to the phosphate value chain through the cascading usage of by-products.

3.1.3 Case C: Alternative Sources of Biomass for a New Biorefinery Concept

This case study derives from the BioSC research project "Advanced pulping for perennial plants: A holistic and sustainable integrated lignocellulose biorefinery concept" (AP3), the aim of which is to establish a novel biorefining process that uses perennial biomass grown on marginal soils and fractionated by the OrganoCat technology (Grande et al., 2015). Such a comprehensive approach would require that actors from different fields (agriculture and forestry, pulp and paper, chemicals, energy, and biotechnology) become interlinked in a new manner (Fig. 3).

The underlying goal of this novel technology is to offer an alternative to contemporary biorefineries, which use predominately food crops such as corn, sugar beet, or rapeseed. The development of these facilities, often referred to as first-generation

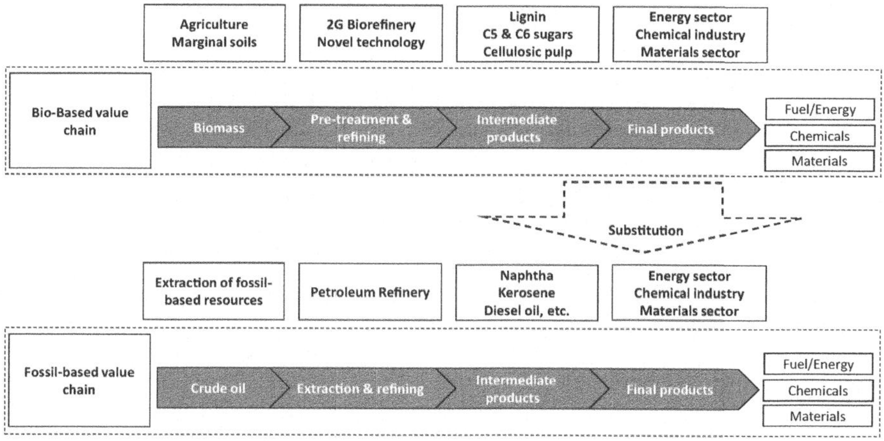

Fig. 3 Lignocellulosic biorefinery value chain showing the substitution effect of the fossil-based value chain. Source: own adaptation from https://www.biosc.de/AP3_Innovationsmanagement

biorefineries, has led to land use competition between food, fuel, and fiber. A promising solution to overcome such issues is to integrate non-food perennial plants (secondary biomass) that can grow on lands with little agricultural value (due to poor soil quality, poor water supply, challenging terrain, or other reasons) into biomass production.

The AP3 project explored three potential plant species—*Miscanthus*, *Sida*, and *Silphium*—which can be fractionated into lignin C5 and C6 sugars and cellulosic pulp using an efficient processing technology. These products can then be treated to obtain chemicals, bio-based materials (e.g., adhesives), and ethanol (Baruah et al., 2018; Dahmen et al., 2019). Although competition with first-generation biorefineries is challenging due to ethanol's more cost-efficient production, the true potential of lignocellulosic biorefineries lies in the valorization of the other intermediate products, especially lignin. As an organic polymer, its building blocks can be extremely useful for producing other chemicals and materials. Data from desk research on patent databases have indicated three main potential applications for lignin derived from second-generation biorefineries: (a) jet fuels and fuels for maritime transport; (b) macromolecules (e.g., lignin in adhesives, binders, and carbon fibers); and (c) aromatic compounds. These applications represent much higher-value offerings than ethanol and could form a revenue backbone to justify the higher processing costs of lignocellulosic biorefineries.

Analogous to the previous case studies, the successful commercialization of this holistic biorefinery concept would generate a substitute value chain in which previously separate actors establish new relationships beyond their respective industries. An agricultural producer, for instance, could become a raw material supplier for the chemical industry, substituting agricultural products for the petroleum-based components of the existing value chain.

3.2 Source of Data and Method of Analysis

Data for the three cases were obtained from interviews with expert value chain actors, all located in Germany, carried out during the abovementioned research projects. The farmers and firms interviewed did not participate directly in the projects but were chosen according to their reputations and market positions. Through these expert interviews, we explored the challenges arising when a new value chain emerges and contrasted them with evidence from the literature.

We applied a semi-structured method for the interviews, which were carried out face to face, on videocall, or by phone (Table S2). The interviews were organized in three parts (Carraresi et al., 2018): first, the project and the emerging technologies were explained to the interviewees to provide them with the details necessary to form an opinion; second, company information was asked of the interviewee (e.g., current business, customer base, etc.) to form an overview of the company's profile; and third, an open discussion was begun, moderated by key questions about potential organizational changes within the value chain. Here, interviewees needed to foresee the changes involved in implementing the new process within the current value chain and were asked to outline the potential benefits and barriers resulting from this implementation. The interviews were transcribed, hand-coded according to categories, and analyzed.

For the case of extracting secondary metabolites from tomato by-products, we interviewed two tomato growers and four industry representatives (one for the chemical industry, two for the food supplement industry, and one for the cosmetics industry). The interviews ranged in length from 20 to 60 min.

In the case of the release and recovery of phosphorus from rapeseed oil press cakes, we interviewed four managers who were key representatives of the emerging value chain: one phosphate processor, one rapeseed oil processor, one industrial biotech company, and one feeding mix producer. The interviews lasted from 30 to 90 min.

For the case of alternative sources of biomass for a new biorefinery concept, we interviewed eight companies, each covering one or more value chain steps (biomass production, pre-treatment/refining, formulation of intermediate products, and formulation of final products), as well as one researcher, to provide an overview of the chemical industry and lignocellulosic innovation field. These interviews ranged from 20 to 80 min.

4 Results

4.1 Challenges Emerging in Case a: Extraction of Secondary Metabolites from Tomato by-Products

The results of the interviews indicate that farmers and company representatives are generally interested in the novel process of valorizing tomato plant by-products and extracting valuable ingredients. Farmers who are already engaged in activities other than their main business (e.g., farmer-to-consumer direct marketing) see opportunities to further diversify their income sources by providing companies with their green residual biomass. Producers of dietary supplements and cosmetics indicate that they would prefer regional suppliers for their active ingredients, as they are currently unsatisfied with the ingredient quality, transparency, and customer service of foreign producers (e.g., from Asia). Furthermore, company representatives observe an increasing consumer trend of sustainability awareness and health consciousness, which aligns with the benefits of the proposed process. By producing bio-based consumer goods with sustainable and health-promoting ingredients, companies would be able to respond to this trend and add a price premium. Despite several perceived benefits of the technology for valorizing horticultural by-products, farmers and company representatives are also aware of potential obstacles to implementing the novel process (Table 1).

Concerning organizational challenges, both farmers interviewed were relatively open to new farming techniques, but they were also concerned about more conservative farmers who could be hesitant to adopt novel practices that would enable by-product valorization. Farms with a strategic focus on cost-leadership (Porter, 1980), in particular, might be focused on their main business (i.e., tomato production) and refuse to differentiate themselves, aiming instead to achieve operational excellence (Treacy & Wiersema, 1993). Such farmers tend to be path dependent and are seldom willing to change their business model.

Regulatory challenges relate to compliance with strict regulations and standards for the process of isolating secondary metabolites (e.g., depending on the target industry, only specific solvents are allowed). Respondents indicated compliance with regulations to be a potential challenge. Regulation is also relevant to the commercialization of the final product. Indeed, representatives of the cosmetics and dietary supplement industries highlighted the importance of communicating the beneficial health effects of secondary metabolites to consumers. However, complying with the current Nutrition and Health Claims Regulation (EC) No. 1924/2006 would require high R&D expenditures to establish health claims for food and cosmetics ingredients, which companies are not always willing to invest (Khedkar et al., 2017).

Economic challenges are the most numerous. Although horticultural by-products are currently underutilized, farmers would expect financial incentives for harvesting and drying the biomass and supplying it to chemical companies. Consequently, these costs must be considered in the economic evaluation of the whole process (Bloomberg New Energy Finance, 2012). Additionally, company representatives foresaw intense competition from large-scale plant extract companies in Asia. Along with the availability of several substitute products for rutin and solanesol, these issues entail high barriers to entry into the market for the active ingredients, which could eventually be overcome by applying a cost leadership or focus strategy (Porter, 1980). Finally, company representatives questioned the profitability of the whole process.

Table 1 Comparison of challenges revealed within the case studies

	Challenges mentioned by case companies		
	Case A—Extraction of secondary metabolites from tomato by-products	Case B—Phosphorus release from rapeseed oil press-cakes	Case C—Alternative biomass for a new biorefinery concept
Organizational challenges			
Dominance of existing business models	X	X	X
Lack of access to appropriate networks		X	X
Lack of competencies/ infrastructures		X	X
Regulatory challenges			
Compliance with standards and regulations	X		
Lack of industry standards		X	X
Economic challenges			
Lack of financial incentives	X		
Intense competition with conventional products/high entry barriers	X		
Economic profitability/ process efficiency	X	X	X
Pricing issues and bargaining power along the chain		X	X
Long timespan to get returns/high opportunity costs			X
Low technology readiness level		X	X
Geographical challenges			
Environmental sustainability efficacy	X		X
Distance between locations of chain steps			X
Product quality challenges			
Heterogeneity in the raw materials/uncertain steady supply	X	X	X

Source: own illustration

As farmers and chemical companies are often distantly located from each other, geographical challenges also emerge. High costs for the transport of biomass and consequent high CO2 emissions might pose problems in terms of profitability and environmental sustainability.

With reference to the challenges associated with the quality of the final products, representatives of chemical companies expressed concerns about receiving a steady supply of biomass from farmers, who might not be able to continuously provide green tomato by-products for component extraction. These companies rely on a regular supply of biomass to guarantee the delivery of active ingredients to their own customers.

4.2 Challenges Emerging in Case B: Phosphorus Release and Recovery from Rapeseed Oil Press Cakes

The extreme novelty of the technology and the lack of process upscale may lead respondents to underscore mainly the obstacles and bottlenecks of its implementation, to the detriment of potential advantages inherent in this bio-based innovation. There are undoubtedly challenges to tackle within the value chain in adopting this innovative phosphate recovery technology and diffusing it on the market, but there would also be incentives, of which respondents seem not to be aware at present (Carraresi et al., 2018).

Organizational challenges seem to predominate in this case, as they cause cascading effects that also hamper the implementation of this clean technology. It is difficult to integrate the new technology into the value chain due to the lack of networking among chain actors and cross-sectoral relationships, as respondents pointed out. The value chain is newly emerging, and the companies potentially involved in the process have never been in contact before and are following different industry logics, making it extremely hard to start a new industry boundary spanning process together. All the interviewees raised the fact that, on one side, the sector of rapeseed oil press cakes is fragmented and, on the other side, it would be necessary to change the customer base for the press cakes. This leads to another issue that makes implementing the new technology in the value chain difficult: companies would have to change their business models based on who is willing to invest or has the competencies to run the bioreactor needed for the extraction process. The results of the interviews reveal that value chain actors struggle with such disruptions and are not yet ready for the new business model. The results also underline the lack of necessary infrastructure and competencies, which also justifies actors' skepticism toward investing in the new technology. To make this innovation work, investment in a bioreactor and the acquisition of relevant bio-technology competencies are needed.

Furthermore, this technology might face regulatory challenges. The new variant of the enzyme used to extract phosphate has different properties than conventional ones, which might pose the need for companies to comply with new regulations and industry standards—which do not yet exist—according to their targeted use of the bio-based polyphosphate. This creates uncertainties for the phosphate producer in particular, who has direct access to the polyphosphate market and who highlighted this challenge in the interview.

Regarding economic challenges, the polyphosphate company, which is specialized in formulations tailored to customer requirements, was particularly concerned about the technology's development status. The technology needs yeast fermentation data to define the functionalities of the bio-based polyphosphate. At present, these insights are still only exploratory, indicating a low technology readiness level. This low readiness also entails unpredictability related to the integration of complementary technologies, such as the enzyme variant formulation, bioprocess, and polyphosphate production, which need to be synchronized to ensure the required phosphate productivity. The rapeseed oil processor also underlined that the sale of press cakes to phosphate processors instead of the feeding industry might create challenges for the pricing process associated with a different level of bargaining power between supplier and buyer. Although rapeseed oil processors could obtain a higher value added for their press cakes, resulting in greater power in contract relationships, it would be difficult for them to sell the press cakes at a higher price, at least in Germany, where there is no "quota system" like in the Netherlands (USDA, 2017). If there were such a quota system, which regulates the phosphate content in feeding mix, breeders would likely also be willing to pay a premium price for feeding mix lower in phosphate content.

The same difficulties associated with a lack of networks and changes in the customer base also hold for companies that must develop the new enzyme formulation. The efficiency and profitability of the whole process were also mentioned by all respondents as potential challenges, as there is a general fear among chain actors that the quantity of rapeseed press-cakes necessary to obtain enough polyphosphate with the novel technology would be fairly high, due to both the proportion of phosphate in the press cakes (around 1%) and the release capacity, which is currently limited. Therefore, the interviewees shared the opinion that bio-based polyphosphate would represent only a niche production. Nonetheless, according to the estimates of the German Federal Ministry of Food and Agriculture (BMEL), the current number of press cakes produced would cover the polyphosphate demand (BMEL, 2014; Carraresi et al., 2018).

Geographical challenges were not reported by the interviewees, but they did underline challenges associated with the quality of the final product. The abovementioned lack of access to networks also leads to difficulties surrounding integration along the value chain. The quantity of oil press cake producers makes it problematic to define contracts for the procurement of a steady supply of raw materials with specific characteristics. The need to obtain a specific quantity of press cakes to recover the amount of phosphate that would satisfy the demand requires the collection of raw material from different suppliers, resulting in heterogeneity in the chemical properties of the press cakes (Simon et al., 2016).

4.3 Challenges Emerging in Case C: Alternative Sources of Biomass for a New Biorefinery Concept

Similar to the other featured cases, the core of the new biorefinery concept—its processing technology—is still in a laboratory-scale phase and has not yet been proven successful in commercial conditions. Furthermore, certain high-value applications of the derived products are under investigation in the current research but have not been fully commercialized (e.g., the application of lignin as a building block for aromatic chemicals). An additional complication stems from the dependence of the concept's potential development on farmers' willingness to cultivate and supply new plant species that would require new agricultural practices.

The organizational challenges mentioned by the majority of respondents are associated with difficulties in changing and adapting the current business model, or unwillingness to do so, toward a model enabling the function of a lignocellulosic biorefinery. Companies in the biomass processing sector are focused on their day-to-day operations, and they are not eager to undertake programs for dealing with a new raw material. We also detected a lack of infrastructure and competencies that may hamper the implementation of the new technology. There are incompatibilities between the novel perennial plants and existing agricultural machinery or practices, and managers often lack the innovation capacity and strategic planning to implement the changes needed to integrate biorefinery processes. Difficulties are also associated with the lack of access to appropriate networks, as companies involved in the potential value chain belong to separate industries. For example, both cosmetics and food companies would be interested in lignin-derived products, but they might not be aware of where to obtain lignin: from pulp and paper companies. Therefore, we observe a missing link between actors in the emerging value chain and between research and industry, which is unaware of the new technology available.

Regulatory challenges were also underlined. Both the specialized literature and our respondents drew attention to the weak policy support for bio-based products other than fuel. The European Union directives on renewable energy and fuel qualities have set no targets for chemicals, materials, or other biorefinery-derived products and in so doing have pushed the research effort and public discussion in a single direction. There is also a lack of EU policies on biorefineries and their associated value chains (Bauer et al., 2017). In the USA, in contrast, deliberate efforts have been made to identify targets for both biofuel and bioproducts, as well as to create a market environment for them. Our respondents expressed optimism that EU regulations may develop in a similar direction.

Several economic challenges were also revealed by the respondents. First, the novel lignocellulosic biorefinery value chain encounters issues stemming from the very source of its feedstock: perennial plants characterized by long establishment periods. In the case of Miscanthus, for instance, it can take up to 4 years before the first harvest (Caslin et al., 2010). Combined with uncertain markets and unstable prices for this type of biomass, this implies high opportunity costs for farmers in the first years of their involvement in this process. Also, the low readiness level of technologies to valorize lignin and the lack of development in application areas beyond the scientific proof of concept would require a long, cost-intensive process to turn these proofs of concept into marketable products, with a consequent long time span before getting returns.

The uncertain economic viability and profitability of the novel process pose another barrier. Not only would the biorefinery need to optimize the collection, transportation, and storage costs of biomass, but its products would also have to be economically competitive with their fossil-based equivalents and the compounds derived from feedstock that is also a food source (such as first-generation ethanol from corn, sugar beet, or rapeseed). Because ethanol is a high-volume/low-value product, processing technologies must achieve reduced production costs for ligno-cellulosic ethanol to merit a shift in the business model. In the same vein, industrial applications for high-value lignin are still niche markets, so the volumes of final compounds are still very low. Also, the seed market for *Miscanthus*, *Sida*, and *Silphium* is not yet highly developed. Most respondents mentioned pricing issues at both the business-to-business and business-to-consumer levels. Companies producing final product applications are sensitive to price and unwilling to pay a premium price for a bio-based ingredient with the same properties as an oil-based one. "It is the opposite—people expect a cheaper solution based on lignin," said a representative of an intermediate product company. The same attitudes had been experienced in the business-to-consumer channel, where final consumers are also unwilling to pay a price premium for bio-based products. These challenges are connected with the lack of differentiation between biomass sources at the stage of intermediate or final products. This is especially true for chemicals and materials, for which the attributes of "biodegradability" or "bio-based content" are communicated but not the origin of the biomass. This lack of communication does not help increase the perceived value or final users' willingness to pay.

Geographical challenges in this case are related to the distance between chain actors. Key suppliers were found to often be located far from each other, which constitutes a problem because the biomass is heavy and expensive to transport (unlike liquid petroleum). This implies, in turn, issues for the environmental and economic sustainability of the process.

Product quality challenges for this technology are associated with the complex composition of biomass, which can be fractionated into its main components and subsequently processed into a variety of products. Unlike the petroleum feedstock currently in use, biomass possesses a high degree of functionality and requires unique reaction conditions (FitzPatrick et al., 2010). This variability of plant-based feedstock leads to heterogeneous biorefinery products (e.g., lignin) and poses a significant technical challenge for value chain partners attempting to produce lignin-derived chemicals of high purity, in particular (Beckham et al., 2016). Respondents pointed out that biomass composition varies not just from source to source but also from harvest to harvest, meaning that different weather conditions or agricultural practices can result in desirable (or undesirable) characteristics in the cultivated plants. The optimal condition and composition of biomass for a given final product is still uncertain; this knowledge can be obtained only through tight cooperation between all members of the novel value chain (e.g., from farmers and suppliers of agricultural inputs, through to bioengineers, to the manufacturers of bio-based chemicals and materials). Existing research on matching the properties of bio-based materials with those of existing oil-based ones is limited. A novel, independent standard for bio-based products seems necessary to establish these by means of defining an own product category.

5 Discussion

In line with the objectives of this chapter, the review of the three case studies has underlined several challenges facing value chain actors in implementing emerging bio-based technologies. Some of these challenges confirm what the literature on bioeconomy has already noted, whereas others are peculiar to the cases analyzed.

Concerning organizational challenges, the dominance of existing business models was revealed to be the biggest hurdle in all three case studies (Carraresi & Bröring, 2021; Golembiewski et al., 2015; Johnson et al., 2008; Aid et al., 2017). Companies and farmers showed themselves to be resistant to change, preferring to maintain their current strategy and aiming to achieve operational excellence (Jernström et al., 2017; Treacy & Wiersema, 1993), even if this might lead to certain path dependence (Sydow et al., 2009; Sydow et al., 2020). They are hesitant to implement novel technologies—and potentially valuable ones, though this remains uncertain—when they have already obtained core competence in a well-established process. The dominance of extant business models is associated mainly with the lower profitability of the new bio-based technologies than of the conventional fossil ones—which was underlined as a further barrier. The hesitance to modify the business model may also be connected to the initial low returns that companies would earn. Normally, bio-based innovations require time to be fine-tuned and scaled up; therefore, companies need to take a long-term perspective to see potential future advantages and returns (Bocken et al., 2014), which most fail to do. Moreover, due to knowledge spillover, the first investment in a bio-based innovation normally generates lower financial returns than expected. Spillover implies that an innovation will be better exploited over time by its recipient firms, while knowledge is gradually integrated and absorbed to the detriment of the first mover that incurred the full R&D costs (Dechezleprêtre et al., 2017; Stern & Valero, 2021). Therefore, bio-based innovations are often seen as risky projects characterized by market failure.

Farmers and processors are also rooted in their organizational routines and culture, which makes them averse to change and afraid to explore new business ideas if they do not see a concrete opportunity (Conti et al., 2021; Kirchherr et al., 2018). Previous literature on organizational inertia demonstrates that it hampers business model innovation (Moradi et al., 2021; Huang et al., 2013). However, when profits and results are uncertain, as in the case of emerging bio-based technologies, companies are not inclined to modify their business models without concrete incentives (Moradi et al., 2021; Stern & Valero, 2021), especially if their current business model is performing successfully (Chesbrough, 2003). For example, in the cases of secondary metabolite extraction (case A) and phosphate recovery (case B), new chain relationships and networks would have to be established to make the bio-based technology work (Aid et al., 2017), whereas in the case of the new biorefinery concept (case C), farmers would have to shift their cultivation plan or add to it the appropriate perennial plants, which are characterized by a long establishment period (Caslin et al., 2010). Therefore, companies would need to be able to afford high opportunity costs to switch to new infrastructure assets or new biomass sources, as the case may be.

Regarding regulatory challenges, there could be a regulatory-push effect if targeted regulation, policy support, and financial incentives were introduced for bio-based products (Horbach, 2008; Rehfeld et al., 2007). For instance, if a threshold limit on the phosphate content of slurry were regulated by a specific policy, companies would be better incentivized to engage in a process that produces feeding mix naturally low in phosphate, like the rapeseed press cakes in the case at hand. However, the lack of industry standards and norms specifically dedicated to governing bio-based products represents an obstacle to technology implementation, as it causes uncertainty about the features required in raw materials, ingredients, and final products to receive approval (Golembiewski et al., 2015).

Finally, from an economic viewpoint, despite the potential advantages of innovative bio-based technologies, some concerns associated with their overall profitability and the quality of the final industrial and consumer products remain. First, companies need to achieve a competitive advantage over the producers of conventional fossil-based products; thus, they require a higher technological readiness level and proven efficiency of new technologies to understand the time horizon at which they will achieve profits and to adapt their strategic planning accordingly. As Müller and Laibach (2022) assert, the extracted products must be valuable and expensive, or their content in the biomass must be high, for the process to be profitable after the costs of investment are compensated. Second, final products of an appropriate quality and in a sufficient quantity to cover the demand must be produced. Obtaining highly pure compounds is one of the main challenges due to the mixed nature of the residues coming either from different suppliers or sources of waste and thus requires significant investment (Donner et al., 2020). Therefore, processors require the assurance of a steady supply and reasonable expectation of feedstock and raw materials with homogeneous properties (Müller and Laibach, 2022; Simon et al., 2016). This would allow processors to meet the requirements of their target customers and final users or consumers. As these emerging bio-based technologies are currently in a laboratory phase, it is difficult to determine the exact amount of feedstock and raw materials needed to cover the demand. Therefore, the availability of biomass must also be considered in the scaling-up process to foster a willingness to adopt such technology.

Based on the challenges underlined in this study, some recommendations can be derived. First, clustering the biomass supply in a few "collection centers" and generating specialized geographical areas in which the entire new value chain could be located, such as districts or agro-parks (Smeets, 2011; Wubben &

Isakhanyan, 2011), might ease the required networking and the supply of raw materials (Donner et al., 2020; Carraresi et al., 2018). In this way, chain actors would be located close together and easily reachable, contributing to better control of the biomass available and its properties. This would reduce the costs and emissions associated with the transport of bulky biomass and positively impact local economies through the valorization of local resources, the generation of job opportunities, and the inclusion of micro and small farmers and companies. Second, extending these districts to include academics and research institutions might help close the gap between research and industry by facilitating public and private partnerships. This would foster the implementation of emerging bio-based technologies and encourage the establishment of sustainable ecosystems (Binz et al., 2014; Markard et al., 2012). Third, involving companies in the co-creation of bio-based technology development and ensuring that it follows user-centric processes could facilitate customized solutions, reduce user risk perception, increase adoption, and reduce market failure. For example, living labs (Bergvall-Kåreborn & Ståhlbröst, 2009) could be a valuable option for engaging researchers, firms, and, at a later stage, governmental institutions and stakeholders of emerging technologies in the entire process of innovation management until validation in real-world environments (Almirall & Wareham, 2011).

Another leverage point would be for research organizations to carry out communication and outreach activities aimed to inform, raise awareness, and attract the interest of processors, entrepreneurs, policy makers, and other stakeholders regarding the environmental benefits of emerging technologies. It has been shown that despite the challenges of implementing emerging bio-based technologies, managers and farmers who are sensitive to environmental issues and proactive in solving them are willing and eager to adopt sustainable practices, more easily overcoming barriers (Bhatia, 2021; Wensing et al., 2019).

Policy makers would also play a role in triggering the implementation of bio-based processes by introducing directives to regulate chemicals, materials, and other products derived from new biorefinery concepts and processes. They could also recommend that subsidies be offered to farmers who are implementing responsible innovations and generally fostering a circular economy. Due to the diverse challenges in this area, a coordinated and regularly updated package of policies considering different emerging technologies would help overcome market failures and path-dependent behaviors and move instead toward clean innovations (Stern & Valero, 2021).

6 Conclusions

The results of the expert interviews revealed several barriers that still hamper the valorization of agricultural by-products and, more generally, bio-based technologies intended to replace conventional fossil-based ones. Interviewees showed a widespread hesitation to change their current business models and implement the novel bio-based technologies. This hesitation was associated mainly with the uncertain level of profit achievable and the long time span until the overall process could be expected to generate substantial earnings. Regulatory issues and concerns regarding product quality were also pointed out.

Companies' reluctance to reinvent their business models is linked to their lack of absorptive capacity to acquire competencies from hitherto separate industries; this capacity is needed for the implementation of such innovative processes as those discussed in this chapter (Cohen & Levinthal, 1990). This deficiency might lead to a general fear of the new and a tendency to pursue traditional strategic paths. A recommendation for companies would be to open up their organizations and to dialogue with other industries in an attempt to explore potential novel opportunities, such as those offered by research and development in the bioeconomy domain. Companies would need to develop a more ambidextrous capacity (Tushman and O'Reilly, 1996), enabling them to explore new technological opportunities and market spaces while continuing to exploit their core business. This would minimize their risk of failure but also enhance their innovation and movement toward circularity. Therefore, companies should attempt to investigate more widely emerging business opportunities and find new cross-industry possibilities for cooperation to unlock their rigidity, acquire new competencies and know-how, and become less hesitant to adopt emerging bio-based technologies (Bocken et al., 2014; Lozano, 2007; Witjes & Lozano, 2016).

Despite these hurdles, the case studies analyzed also present drivers that outline a promising future for agricultural by-products and alternative biomass sources. First, in these cases, agricultural by-products acquire a higher value added, creating the possibility for suppliers to achieve better bargaining power with their customers. Second, the alternative and cascading usage of agricultural by-products gives suppliers the opportunity to further differentiate their production and obtain premium prices by underscoring their products' bio-based sources and leveraging their positive green image (Carus et al., 2018). Third, processors located downstream in the value chain could also benefit from the bio-based origin of the raw materials and try to segment their market to reach final consumers who are more sensitive to environmental issues. Fourth, although lignocellulosic biorefinery plants have a long establishment period, once established, they can remain in situ for at least 15 years, providing annual harvest and income. The cultivation of perennial plants in marginal soils could be a strategy for farmers to diversify their revenue sources using fields that are not suitable for other agricultural production (Clifton-Brown et al., 2017).

Overall, despite the presence of some barriers, the advantages of these bio-based technologies, which have also been recognized within their respective case studies, indicate that there is scope for follow-up research aimed at overcoming these hurdles and fostering the realization of sustainable innovations. In particular, the effect of carbon pricing/taxes on the economic viability of the discussed cases might be interesting to study in the future.

The limitations of this study include the relatively small number of respondents, which is suitable only for an exploratory study and review. We wanted to provide an overview of the potential obstacles sensed by farmers and companies that might be interested, in the (hopefully) near future, in these emerging bio-based technologies. Due to the low technology readiness level and embryonal stage of the technologies themselves, it was not possible to obtain detailed data about the technologies (e.g., precise costs of investments, life cycle assessment, efficiency), which would have offered a more tangible scenario.

Future research should aim to provide more insight into the cases at hand and to extend the sample of companies interviewed. This sample should include other potential stakeholders, such as producers' and farmers' associations, consumers, and policy makers. Further, future studies would need to implement a deeper economic and life cycle analysis to better assess the feasibility of these processes and their environmental and economic sustainability to foster their adoption by the companies involved in their respective value chains. Finally, further research would need to assess consumer interest in and acceptance of bio-based products to obtain a more complete overview of the market demand, including the downstream part of the value chain and the marketability of the products.

7 Acknowledgments

The authors thank Joana Wensing and Lora Tsvetanova for their support in data collection and analysis for cases A and C, respectively.

We acknowledge funding by the Ministry of Innovation, Science and Research of the German State of North Rhine-Westphalia in the framework of the Bioeconomy Science Center (BioSC, NRW Strategieprojekt No. 313/323–400–00213).

References

Aghion, P., Hepburn, C., Teytelboym, A., & Zenghelis, D. (2019). Path dependence, innovation and the economics of climate change. In R. Fouquet (Ed.), *Handbook on green growth* (pp. 67–83). Edward Elgar Publishing.

Aid, G., Eklund, M., Anderberg, S., & Baas, L. (2017). Expanding roles for the Swedish waste management sector in inter-organizational resource management. *Resources, Conservation and Recycling, 124*, 85–97.

Alblas, A. A., Peters, K., & Wortmann, J. C. (2014). Fuzzy sustainability incentives in new product development: An empirical exploration of sustainability challenges in manufacturing companies. *International Journal of Operations & Production Management, 34*(4), 513–545.

Almirall, E., & Wareham, J. (2011). Living labs: Arbiters of mid- and ground-level innovation. *Technology Analysis & Strategic Management, 23*(1), 87–102.

Augustin, M. A., Cole, M. B., Ferguson, D., Hazell, N. J. G., & Morle, P. (2021). Perspective article: Towards a new venture science model for transforming food systems. *Global Food Security, 28*, 100481. in press.

Bai, C., Kusi-Sarpong, S., & Sarkis, J. (2017). An implementation path for green information technology systems in the Ghanaian mining industry. *Journal of Cleaner Production, 164*, 1105–1123.

Baruah, J., Nath, B. K., Sharma, R., Kumar, S., Deka, R. C., Baruah, D. C., & Kalita, E. (2018). Recent trends in the pretreatment of lignocellulosic biomass for value-added products. *Frontiers in Energy Research, 6*(141), 1–19.

Bauer, F., Coenen, L., Hansen, T., McCormick, K., & Palgan, Y. V. (2017). Technological innovation systems for biorefineries: A review of the literature. *Biofuels, Bioproducts and Biorefining, 11*(3), 534–548.

Beckham, G. T., Johnson, C. W., Karp, E. M., Salvachúa, D., & Vardon, D. R. (2016). Opportunities and challenges in biological lignin valorization. *Current Opinion in Biotechnology, 42*, 40–53.

Bergvall-Kåreborn, B., & Ståhlbröst, A. (2009). Living lab: An open and citizen-centric approach for innovation. *International Journal of Innovation and Regional Development, 1*(4), 356–370.

Bhatia, M. S. (2021). Green process innovation and operational performance: The role of proactive environment strategy, technological capabilities, and organizational learning. *Business Strategy and the Environment, 30*(7), 2845–2857.

Bierly, P. E., & Chakrabarti, A. (2001). Dynamic knowledge strategies and industry fusion. *International Journal of Manufacturing Technology and Management, 3*(1), 31–48.

Binz, C., Truffer, B., & Coenen, L. (2014). Why space matters in technological innovation systems—Mapping global knowledge dynamics of membrane bioreactor technology. *Research Policy, 43*(1), 138–155.

Bloomberg New Energy Finance (2012) Next-generation ethanol and biochemicals: what's in it for Europe? https://www.novozymes.com/en/-/media/Novozymes/en/sustainability/transparency-and-accountability/Documents/2010-Bloomberg-Next-generation-ethanol-and-biochemicals-whats-in-it-for-Europe.pdf?la=en.

BMEL. (2014). http://berichte.bmelv-statistik.de/SJT-3090900–0000.pdf

Bocken, N. M. P., Short, S. W., Rana, P., & Evans, S. (2014). A literature and practice review to develop sustainable business model archetypes. *Journal of Cleaner Production, 65*, 42–56.

Boehlje, M., & Broering, S. (2011). The increasing multifunctionality of agricultural raw materials: Three dilemmas for innovation and adoption. *International Food and Agribusiness Management Review, 14*, 1–16.

Bröring, S., & Cloutier, L. M. (2008). Value-creation in new product development within converging value chains. An analysis in the functional foods and nutraceutical industry. *British Food Journal, 110*(1), 76–97.

Bröring, S., Laibach, N., & Wustmans, M. (2020). Innovation types in the bioeconomy. *Journal of Cleaner Production., 266*, 121939. in press.

Carraresi, L., Berg, S., & Bröring, S. (2018). Emerging value chains within the bioeconomy: Structural changes in the case of phosphate recovery. *Journal of Cleaner Production, 183*, 87–101.

Carraresi, L., & Bröring, S. (2021). How does business model redesign foster resilience in emerging circular value chains? *Journal of Cleaner Production, 289*, Article 125823.

Carus, M., Partanen, A., & Dammer, L. (2018). *Detailed evaluation of green premium prices for bio-based products along the value chain*. Nova Institute for Energy and Innovation.

Caslin, B., Finnan, J., & Easson, L. (2010). *Miscanthus best practice guidelines*. Agriculture and Food Development Authority, Teagasc, and Agri-Food and Bioscience Institute.

Chemat, F., Vian, M. A., Fabiano-Tixier, A. S., Nutrizio, M., Jambrak, A. R., Munekata, P. E., Lorenzo, J. M., Barba, F. J., Binello, A., & Cravotto, G. (2020). A review of sustainable and intensified techniques for extraction of food and natural products. *Green Chemistry, 22*(8), 2325–2353.

Cherubini, F. (2010). The biorefinery concept: Using biomass instead of oil for producing energy and chemicals. *Energy Conversion and Management, 51*(7), 1412–1421.

Chesbrough, H. (2003). The logic of open innovation: Managing intellectual property. *California Management Review, 45*(3), 33–58.

Chua, L. S. (2013). A review on plant-based rutin extraction methods and its pharmacological activities. *Journal of Ethnopharmacology, 150*(3), 805–817.

Clift, R. (1995). Clean technology—An introduction. *Journal of Chemical Technology & Biotechnology, 62*(4), 321–326.

Clifton-Brown, J., Hastings, A., Mos, M., McCalmont, J. P., Ashman, C., Awty-Carroll, D., & Scurlock, J. (2017). Progress in upscaling miscanthus biomass production for the European bio-economy with seed-based hybrids. *Global Change Biology Bioenergy, 9*(1), 6–17.

Clomburg, J. M., Crumbley, A. M., & Gonzalez, R. (2017). Industrial biomanufacturing: The future of chemical production. *Science, 355*(6320), aag0804.

Cohen, W. M., & Levinthal, D. A. (1990). Absorptive capacity: A new perspective on learning and innovation. *Administrative Science Quarterly, 35*(1), 128–152.

Conti, C., Zanello, G., & Hall, A. (2021). Why are Agri-food systems resistant to new directions of change? A systematic review. *Global Food Security, 31*, 100576. in press.

Dahmen, N., Lewandowski, I., Zibek, S., & Weidtmann, A. (2019). Integrated lignocellulosic value chains in a growing bioeconomy: Status quo and perspectives. *GCB Bioenergy, 11*(1), 107–117.

Dangelico, R. M., & Pontrandolfo, P. (2015). Being "green and competitive": The impact of environmental actions and collaborations on firm performance. *Business Strategy and the Environment, 24*(6), 413–430.

Dechezleprêtre, A, Martin, R, Mohnen, M (2017) Knowledge spillovers from clean technologies: a patent citation analysis. Grantham Research Institute on Climate Change and the Environment Working Paper 135: 1–111.

Dinesh, D., Hegger, D. L. T., Klerkx, L., Vervoort, J., Campbell, B. M., & Driessen, P. P. J. (2021). Enacting theories of change for food systems transformation under climate change. *Global Food Security, 31*, 100583. in press.

Donner, M., Gohier, R., & de Vries, H. (2020). A new circular business model typology for creating value from agro-waste. *Science of the Total Environment, 716*, 137065.

Driessen, P. H., Hillebrand, B., Kok, R. A. W., & Verhallen, T. M. M. (2013). Green new product development: The pivotal role of product greenness. *IEEE Transactions on Engineering Management, 60*(2), 315–326.

Eisenhardt, K. M., & Graebner, M. E. (2007). Theory building from cases: Opportunities and challenges. *Academy of Management Journal, 50*(1), 25–32.

Ekman, A., Campos, M., Lindahl, S., Co, M., Börjesson, P., Karlsson, E. N., & Turner, C. (2013). Bioresource utilisation by sustainable technologies in new value-added biorefinery concepts—Two case studies from food and forest industry. *Journal of Cleaner Production, 57*, 46–58.

El Bilali, H. (2019). Research on agro-food sustainability transitions: A systematic review of research themes and an analysis of research gaps. *Journal of Cleaner Production, 221*, 353–364.

FitzPatrick, M., Champagne, P., Cunningham, M. F., & Whitney, R. A. (2010). A biorefinery processing perspective: Treatment of lignocellulosic materials for the production of value-added products. *Bioresource Technology, 101*(23), 8915–8922.

Geroski, P. A. (2000). Models of technology diffusion. *Research Policy, 29*(4), 603–625.

Glaser, B. G., & Strauss, A. L. (1967). *The discovery of grounded theory. Strategies for qualitative research*. Aldine Publishing Company.

Golembiewski, B., Sick, N., & Bröring, S. (2015). The emerging research landscape on bioeconomy: What has been done so far and what is essential from a technology and innovation management perspective? *Innovative Food Science and Emerging Technology, 29*, 308–3017.

Grande, P. M., Viell, J., Theyssen, N., Marquardt, W., de María, P. D., & Leitner, W. (2015). Fractionation of lignocellulosic biomass using the OrganoCat process. *Green Chemistry, 17*(6), 3533–3539.

Horbach, J. (2008). Determinants of environmental innovation e new evidence from German panel data sources. *Research Policy, 37*, 163–173.

Huang, H. C., Lai, M. C., Lin, L. H., & Chen, C. T. (2013). Overcoming organizational inertia to strengthen business model innovation: An open innovation perspective. *Journal of Organizational Change Management, 26*(6), 977–1002.

Jernström, E., Karvonen, V., Kässi, T., Kraslawski, A., & Hallikas, J. (2017). The main factors affecting the entry of SMEs into bio-based industry. *Journal of Cleaner Production, 141*, 1–10.

Johnson, M. W., Christensen, C. C., & Kagermann, H. (2008). Reinventing your business model. *Harvard Business Review, 86*(12), 50–59.

Junker-Frohn, L., Lück, M., Schmittgen, S., Wensing, L., Carraresi, L., Thiele, B., Groher, T., Reimer, J., Bröring, S., Noga, G., Jupke, A., Schurr, U., Usadel, B., Wiese-Klinkenberg, A., & Wormit, A. (2019). Tomato's green gold: Bioeconomy potential of residual tomato leaf biomass as a novel source for the secondary metabolite rutin. *ACS Omega, 4*(21), 19071–19080.

Kardung, M., Cingiz, K., Costenoble, O., Delahaye, R., Heijman, W., Lovric, M., van Leeuwen, M., M'Barek, R., van Meijl, H., Piotrowski, S., Ronzon, T., Sauer, J., Verhoog, D., Verkerk, P. J., Vrachioli, M., Wesseler, J., & Zhu, B. X. (2021). Development of the circular bioeconomy: Drivers and indicators. *Sustainability, 13*(1), 413. in press.

Khedkar, S., Bröring, S., & Ciliberti, S. (2017). Exploring the nutrition and health claims regulation (EC) no. 1924/2006: What is the impact on innovation in the EU food sector? *International Journal of Food Sciences and Nutrition, 68*, 10–17.

Kirby, J., & Keasling, J. D. (2009). Biosynthesis of plant isoprenoids: Perspectives for microbial engineering. *Annual Review of Plant Biology, 60*, 335–355.

Kirchherr, J., Piscicelli, L., Bour, R., Kostense-Smit, E., Muller, J., Huibrechtse-Truijens, A., & Hekkert, M. (2018). Barriers to the circular economy: Evidence from the European Union (EU). *Ecological Economics, 150*, 264–272.

Laudien, S. M., & Daxbock, B. (2016). Path dependence as a barrier to business model change in manufacturing firms: Insights from a multiple-case study. *Journal of Business Economics, 86*, 611–645.

Leeuwis, C., Boogaard, B. K., & Atta-Krah, K. (2021). How food systems change (or not): Governance implications for system transformation processes. *Food Security, 13*, 761–780.

Lozano, R. (2007). Collaboration as a pathway for sustainability. *Sustainable Development, 15*(6), 370–381.

Maaß, O., Grundmann, P., & Polach, C. V. B. (2014). Added-value from innovative value chains by establishing nutrient cycles via struvite. *Resources, Conservation and Recycling, 87*, 126–136.

Maity, S. K. (2015). Opportunities, recent trends and challenges of integrated biorefinery: Part I. *Renewable and Sustainable Energy Reviews, 43*, 1427–1445.

Markard, J., Raven, R., & Truffer, B. (2012). Sustainability transitions: An emerging field of research and its prospects. *Research Policy, 41*(6), 955–967.

Marvel, M. R., Wolfe, M. T., & Kuratko, D. F. (2020). Escaping the knowledge corridor: How founder human capital and founder coachability impacts product innovation in new ventures. *Journal of Business Venturing, 35*(6), 106060. in press.

McCormick, K., & Kautto, N. (2013). The bioeconomy in Europe: An overview. *Sustainability, 5*, 2589–2608.

Michael L., Tushman Charles A., O'Reilly (1996) Ambidextrous Organizations: Managing Evolutionary and Revolutionary Change. California Management Review 38(4) 8-29 10.2307/41165852

Mohan, S. V., Nikhil, G. N., Chiranjeevi, P., Reddy, C. N., Rohit, M. V., Kumar, A. N., & Sarkar, O. (2016). Waste biorefinery models towards sustainable circular bioeconomy: Critical review and future perspectives. *Bioresource Technology, 215*, 2–12.

Moradi, E., Jafari, S. M., Doorbash, Z. M., & Mirzaei, A. (2021). Impact of organizational inertia on business model innovation, open innovation and corporate performance. *Asia Pacific Management Review, 26*(4), 171–179.

Müller, B., & Laibach, N. (2022). Extraction of valuable components from waste biomass. In S. Smetana et al. (Eds.), *Waste to food: Returning nutrients to the food chain* (pp. 147–168). Wageningen Academic Publishers.

Newton, J. E., Nettle, R., & Pryce, J. E. (2020). Farming smarter with big data: Insights from the case of Australia's national dairy herd milk recording scheme. *Agricultural Systems, 181*, 102811. in press.

Nooteboom, B. (2000). Learning by interaction: Absorptive capacity, cognitive distance and governance. *Journal of Management and Governance, 4*(1–2), 69–92.

OECD. (2014). *Biobased chemicals and bioplastics: Finding the right policy balance*. OECD Publishing.
Parisi, C (2018). Research Brief: Biorefineries distribution in the EU. European Commission—Joint Research Centre, Publications Office of the European Union, Luxembourg, pp. 1–8.
Patterson, J., Schulz, K., Vervoort, J., van der Hel, S., Widerberg, O., Adler, C., Hurlbert, M., Anderton, K., Sethi, M., & Barau, A. (2017). Exploring the governance and politics of transformations towards sustainability. *Environmental Innovation and Societal Transitions, 24*, 1–16.
Peters, K., & Buijs, P. (2022). Strategic ambidexterity in green product innovation: Obstacles and implications. *Business Strategy and the Environment, 31*, 173–193.
Pleissner, D., Qi, Q., Gao, C., Rivero, C. P., Webb, C., Lin, C. S. K., & Venus, J. (2016). Valorization of organic residues for the production of added value chemicals: A contribution to the bio-based economy. *Biochemical Engineering Journal, 116*, 3–16.
Porter, M. E. (1980). *Competitive strategy*. Free Press.
Rehfeld, K., Rennings, K., & Ziegler, A. (2007). Integrated product policy and environmental product innovations: An empirical analysis. *Ecological Economics, 61*(1), 91–100.
Rotolo, D., Hicks, D., & Martin, B. E. (2015). What is an emerging technology? *Research Policy, 44*, 1827–1843.
Scarlat, N., Dallemand, J.-F., Monforti-Ferrario, F., & Nita, V. (2015). The role of biomass and bioenergy in a future bioeconomy: Policies and facts. *Environmental Development, 15*, 3–34.
Schmid, O., Padel, S., & Levidow, L. (2012). The bio-economy concept and knowledge base in a public goods and farmer perspective. *Bio-based and Applied Economics, 1*(1), 47–63.
Sheppard, A. W., Gillespie, I., Hirsch, M., & Begley, C. (2011). Biosecurity and sustainability within the growing global bioeconomy. *Current Opinion in Environmental Sustainability, 3*(1), 4–10.
Shilton, A. N., & Blank, L. M. (2012). Phosphorus biotechnology. *Current Opinion in Biotechnology, 23*(6), 830–832.
Simon, B., Bachtin, K., Kiliç, A., Amor, B., & Weil, M. (2016). Proposal of a framework for scale-up life cycle inventory: A case of nanofibers for lithium iron phosphate cathode applications. *Integrated Environmental Assessment and Management, 12*(3), 465–477.
Smeets, P. J. A. M. (2011). *Expedition agroparks: Research by design into sustainable development and agriculture in network society*. Wageningen Academic Publishers.
Stern, N., & Valero, A. (2021). Innovation, growth and the transition to net-zero emissions. *Research Policy, 50*(9), 104293. In press.
Sydow, J., Schreyögg, G., & Koch, J. (2009). Organizational path dependence: Opening the black box. *Academy of Management Review, 34*(4), 689–709.
Sydow, J., Schreyögg, G., & Koch, J. (2020). On the theory of organizational path dependence: Clarifications, replies to objections, and extensions. *Academy of Management Review, 45*(4), 717–734.
Tushman, M. L., & O'Reilly, C. A., III. (1996). Ambidextrous organizations: Managing evolutionary and revolutionary change. *California Management Review, 38*(4), 8–29.
Treacy, M., & Wiersema, F. (1993). Customer intimacy and other value disciplines. *Harvard Business Review, 71*(1), 84–93.
USDA (2017). New phosphate reduction plan sets limits to Dutch dairy production. Global Agricultural Information Network. GAIN Report Number: NL7006. https://gain.fas.usda.gov/Recent%20GAIN%20Publications/New%20Phosphate%20Reduction%20Plan%20Sets%20Limits%20to%20Dutch%20Dairy%20Production_The%20Hague_Netherlands_2-9-2017.pdf
Wesseler, J., & Zilberman, D. (2021). Biotechnology, bioeconomy, and sustainable life on land. In V. Beckmann (Ed.), *Transitioning to sustainable life on land* (pp. 415–435). MDPI.

Witjes, S., & Lozano, R. (2016). Towards a more circular economy: Proposing a framework linking sustainable public procurement and sustainable business models. *Resources, Conservation and Recycling, 112*, 37–44.
Wubben, E. F., & Isakhanyan, G. (2011). Stakeholder analysis of agroparks. *International Journal on Food System Dynamics, 2*(2), 145–154.
Yan, N., Liu, Y., Gong, D., Du, Y., Zhang, H., & Zhang, Z. (2015). Solanesol: A review of its resources, derivatives, bioactivities, medicinal applications, and biosynthesis. *Phytochemistry Reviews, 14*(3), 403–417.
Yin, R. K. (1994). *Case study research, design and methods* (2nd ed.). Sage Publications.

Building Circular Innovation Ecosystem in Industrial Port Territories: The Case of Dunkirk, France

Aina Ndrianjara Andriamanantena, Blandine Laperche, and Sophie Boutillier

Abstract This chapter highlights the link between the circular economy and the innovation ecosystem in an industrial port territory. We assume that a circular project at the territorial level can lead to the emergence of an innovation ecosystem called "circular innovation ecosystem" which is characterized by its actors, relationships, and resources and results in the creation of value at both micro and meso levels. In applying it to the case of the industrial port territory of Dunkirk in France, we aim to assess, via a set of quantitative and qualitative indicators, the existence of each of the components of a circular innovation ecosystem. Our results show that the circular project in this industrial port territory, oriented toward industrial ecology, has been an impetus for the development of such a circular innovation ecosystem, which is still in its development phase. Our research contributes to the definition and assessment of circular innovation ecosystems, which can be replicated in other industrial port territories.

1 Introduction

Dependence on fossil fuels coupled with the constantly increasing costs of raw materials now more than ever equates to economic vulnerability for enterprises, territories, and/or countries. Ecological awareness has led to the introduction of new environmental standards, increasingly fierce competition, and the heralded scarcity of resources. All these new factors are causing enterprises, territories, and countries to adopt new business models and industrial reconversion approaches. All these new factors are inducing enterprises, territories, and countries to adopt new business models and industrial reconversion approaches that can take the form of circular economy (CE) strategies and policies (Guillard, 2018; WWF, 2014). The CE implies

A. N. Andriamanantena (✉) · B. Laperche · S. Boutillier
Université du Littoral Côte d'Opale (ULCO), Centre for Research on Innovation and Industrial Strategies (ISI), Dunkirk, France
e-mail: aina.njara@univ-littoral.fr; blandine.laperche@univ-littoral.fr; sophie.boutillier@univ-littoral.fr

V. Prokop et al. (eds.), *Business Models for the Circular Economy*, Sustainability and Innovation, https://doi.org/10.1007/978-3-031-08313-6_6

a profound transformation in the traditional way business is carried out, as well as in the way resources are exploited. It is often opposed to the mainstream "linear" economy pattern that has been the theoretical foundation for economic development from the first industrial revolution, based on the "take-make-dispose" practices with no regard for the environment (Blomsma & Brennan, 2017). CE can, instead, be at the origin of a new industrial model (Gallaud & Laperche, 2016). They include new ways of creating value through innovation around the principles of "circularity" (e.g., reuse, industrial ecology, extension of the life cycle of products, and recycling) (Lacy et al., 2020). As an umbrella concept, it can be defined as "a strategy that emerges to oppose the traditional open-ended system, aiming to face the challenge of resource scarcity and waste disposal in a win-win approach with economic and value perspective" (Homrich et al., 2018, p. 534).

The CE is often studied at the macro-level (public policies) and at the microlevel (enterprises); studies also focus on the new business models that they develop (Konietzko et al., 2020; Lopez et al., 2019; Pieroni et al., 2019). It is less the case at the level of territories, and, most of all, the process by which the CE can have a role in territorial development is less studied and is rarely applied to industrial port territories (Veyssière et al., 2021). From this research gap, we are especially interested in industrial port territories and the way in which the CE can constitute a new economic model for their development. Our main goal in this chapter is thus to analyze the mechanisms through which a CE project could contribute to territorial development in industrial port territories, which we study through the emergence of innovation ecosystems. These are made of a set of actors developing close economic relationships in order to generate value at the micro (enterprises) and meso (territories) levels, through new products, services, and organizations, i.e., through innovations (see, e.g., Granstrand & Holgersson, 2020; Gomes et al., 2018; Tsujimoto et al., 2018; Walrave et al., 2018).

Our methodology is based on the development of a theoretical framework explaining how CE projects can contribute to the emergence and strengthening of the main characteristics that underpin and define innovation ecosystems. We then apply this to the case of industrial port territories, and especially to the case of the industrial port territory of Dunkirk, in the north of France,[1] which is the French pioneer of industrial ecology (one of the pillars of the CE) (Kasmi, 2018). We use the

[1]The port of Dunkirk is the third largest French commercial port and is ranked seventh in the North European Range which stretches from Le Havre to Hamburg (Port de Dunkerque, 2021). With 53 Mt. processed on its docks, it is renowned as a large bulk port for its many industrial installations. These installations are mainly heavy industries (steel and metallurgy), energy producers (first nuclear facility in Europe for electricity production), LNG (liquefied natural gas) terminals, and renewable energy producers (biofuels or wind turbines).

Source: Port and Corridors. (2020). Résultats des GPM Français et de leurs principaux concurrents. Trafics Portuaire 2019. Consulted on May 2021, link https://www.umlr.fr/wp-content/uploads/2020/05/R%C3%A9sultats-2019-des-Grands-Ports-Maritimes-fran%C3%A7ais-et-de-leurs-principaux-concurrents-MM.pdfCorridors

relevant indicators allowing us to verify the existence of the main characteristics of an innovation ecosystem in this industrial port territory.

Our work contributes to the definition of the concept of "circular innovation ecosystem" that is to say innovation ecosystems built from a CE project, which can be considered as an analytical tool to understand the mechanisms by which a CE project could contribute to territorial development, especially in industrial port territories. It also presents a methodology for the assessment of the development of such circular innovation ecosystems.

The remainder of the chapter is as follows: based on an analysis of the literature, the first section describes our theoretical model built on the hypothesis that CE projects could be at the origin of innovation ecosystems in industrial port territories. The methodology is described in the second section with a description of the industrial port territory of Dunkirk that we have chosen and then our approach to tackle our case study. Section three presents the results and discusses the notion of circular innovation ecosystems. Section four concludes the chapter.

1.1 The Circular Economy as a New Economic Model for Industrial Port Territories

Well-known experiences of CE (such as industrial ecology) often occur in industrial port territories like the industrial symbiosis (IS) in Kalundborg. In this first part, we propose to revisit the concepts of the circular economy and innovation ecosystem and their potential links with industrial port territories. To do so, we present the relationship between circular economy and industrial port territories, as well as the development of the circular economy in this type of territory, and then we show how a circular economy project can lead the actors of the territory to group together and form a "circular innovation ecosystem."

1.2 The Circular Economy in Industrial Port Territories

1.2.1 The Circular Economy as a Model of Development for Industrial Port Territories

The "circular" economic model is positioned as a recent model for economic development and sustainability by promoting the reuse of waste and closing the cycle of resources (Blomsma & Brennan, 2017). As a recent concept, the CE is about rethinking the way resources are used. Homrich et al. (2018) underline the importance of this use of resources by providing evidence of the scarcity of the available resources and the challenge modern societies face to close the traditional open-ended system by including waste disposal to achieve a win–win approach with an economic and value perspective.

Popularized in the 1990s, the model of the CE is inspired by ideas of industrial ecology in the 1980s through a rethinking of industrial processes (Frosch & Gallopoulos, 1989). Adopted by institutions, the concepts of industrial ecology and CE are widely disseminated in Europe. According to the Ellen MacArthur Foundation (2015), CE finds its foundations in several streams including, among others, industrial ecology. For the French environmental agency, industrial ecology is one of the roots and pillars of the circular economy (Geldron, 2013; Gallaud & Laperche, 2016).

Industrial ecology aims to optimize local management of waste and resources by densifying interactions between stakeholders occupying the same geographical or functional area (Mat, 2015). It takes place at the scale of territories and can be defined as a social construct made by a set of stakeholders rooted in a defined geographical area (Pecqueur & Zimmermann, 2004). As a complex and evolving system, a territory associates a set of actors and the resources that these actors use, develop, and manage in a specific geographical space. (Veyssière et al., 2021). There are different types of territories including industrial territories which are organized around a concentration of various kinds of industries. In this research, we are interested in industrial port territories that can be seen as particular and complex industrial territories due to their geographical location at the convergence of land and sea networks (Foulquier, 2012).

Industrial port territories can be categorized as a combination of three distinctive areas: urban, agricultural, and industrial port (Mat, 2015). Each area has a distinct role and together constitutes an industrial port territory. The industrial port area, also called the industrial port zone, represents the interface area which allows and facilitates the flow of materials and energy to circulate. Having a logistics function (node, transit platform, fuel storage, etc.), this zone also concentrates the industrial fabric of the industrial port territory which is generally connected to the port terminals for straightforward access to gross flows (flow of raw materials, energy flow, etc.). This industrial fabric is often characterized by heavy industries (iron and steel, metallurgy, production of chemical products) which will transform raw flows into products before going through the port for exportation or going into the hinterlands.

1.2.2 Development of Circular Economy in Industrial Port Territories

In Europe, the CE is gaining momentum as one of the main blocks of the European Green Deal,[2] which is Europe's new agenda for sustainable growth. For the European Commission, the transition to a CE is seen as a means to reduce pressure on natural resources and at the same time an opportunity to create sustainable growth and jobs (European Commission, 2020). We can propose a dual relationship between CE and industrial port territories, presented in Fig. 1.

[2]https://ec.europa.eu/info/strategy/priorities-2019-2024/european-green-deal_en

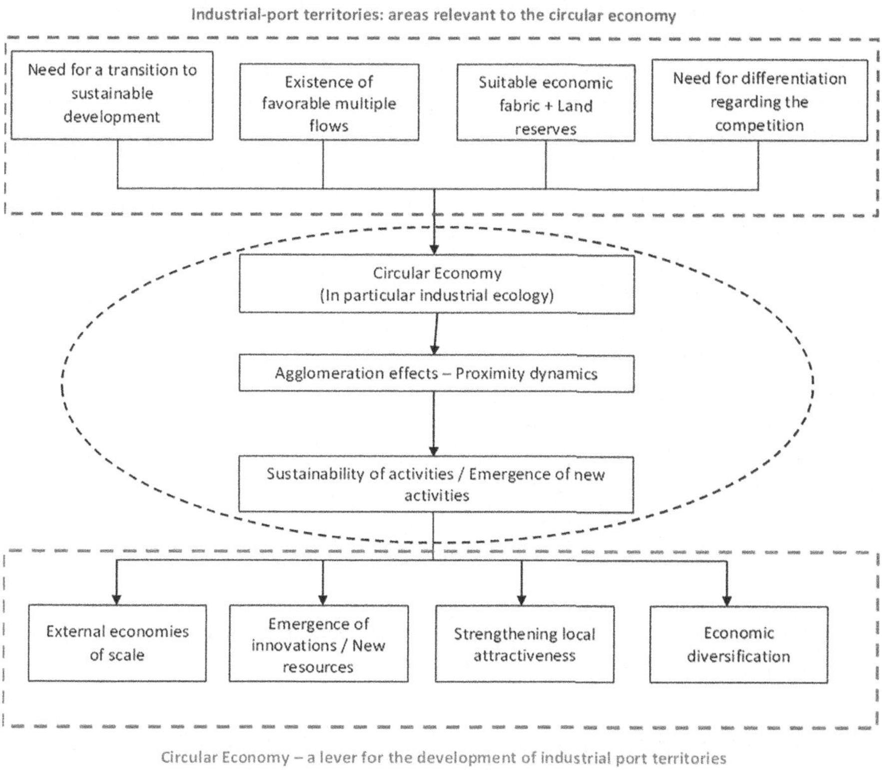

Source: Andriamanantena et al. (2020).

Fig. 1 Circular economy and industrial-port territories. Source: Andriamanantena et al. (2020)

Industrial port territories stand out from the industrial territories by their geographical location associating a land area but also a maritime space (generally by the sea or close to the waterways within a country) which make them gateways to the hinterlands.

These particular territories are more severely affected by the scarcity of raw materials and the rise in the price of fossil fuels due to their "linear" development models. Harsher environmental constraints and competition from other territories lead industrial port territories to find new ways to continue their development. Nevertheless, these concerns present opportunities. As shown in Fig. 1, a dual relation between these particular territories and the CE can be advanced. Industrial port territories are relevant areas for the development of CE activities because of their need to stand out from the competition and also to adopt a more sustainable pathway to continue to develop. The existence of favorable multiple flows (i.e., waste, raw materials, etc.) presents new opportunities to develop new activities related to CE. The dense economic fabric and the land reserve are well suited for industrial ecology and the implementation of new businesses. On the other hand, a

CE project can enhance the relationship between the actors through the exchange of waste traded among enterprises (Baas, 2000; Baas & Huisingh, 2008). Proximity relations can induce positive agglomeration effects, such as a reduction in the consumption of fossil energy and raw materials and the emergence of innovations through the reuse and valorization of the flows. This leads to some competitive advantages in the form of lower prices for the intakes and economies of scale. New resources and interactions between actors mean new possibilities that can enhance the attractiveness of the territory by attracting new investors and enterprises (Kasmi, 2020).

Although the CE and industrial port territories seem to be a good match, our literature review linking the CE and territorial development based on Veyssière et al. (2021) and our findings highlight 39 references. We can classify these articles which link the CE and ports around four main topics:

The issue of the creation of value at the micro (enterprises) and meso (territories) levels from a CE project is rarely dealt with in the literature and this justifies our research. The model of innovation ecosystem can contribute to a better understanding of the mechanisms of this value creation.

1.3 From the Innovation Ecosystem to the Circular Innovation Ecosystem

The origins of research on innovation systems have their antecedents in the 1980s (Freeman, 1982; Lundvall, 1985). The innovation system is first defined at the national level (National innovation system) as "the network of institutions in the public and private sectors whose activities and interactions initiate, import, modify and diffuse new technologies" (Freeman, 1987, p. 1). This concept is used by researchers to demonstrate the importance of close relationships between actors around a common project to generate innovation at a different level.

The model of an ecosystem is used much earlier but not directly in economics and management (Tansley, 1935). It was not until Moore's research in 1993 that the model of a business ecosystem was used outside its initial sphere of biology and environment and applied to management. Moore (1996) proposes that enterprises could be transposed as part of an ecosystem that he presents as a business ecosystem. This business ecosystem consists of networks of interconnected actors who co-evolve by sharing their skills, technologies, and resources around an innovation. Thus, Moore (2006, p. 33) described the business ecosystem as "the collaborative arrangements through which firms combine their individual offerings into a coherent, customer-facing solution" which proposes a network of actors which is built around a large company.

Other researchers have since then developed the term ecosystem around other models such as service ecosystems, platform ecosystem, or an innovation ecosystem, which is a fairly recent concept that takes on its importance, thanks to Adner's work

(2006). Adner and Kapoor (2010, p. 309) suggested that the prefix "eco" in eco-system is "a way of making interdependencies more explicit" between the stakeholders. Nowadays, the model of an innovation ecosystem is applied in various sectors such as primary industries (energy, agriculture), manufacturing industries (aviation, textile, automobile), service industries (health, Internet, information technologies), and high-tech industries (nano-electronics, semiconductors, electronics) (Yaghmaie & Vanhaverbeke, 2019).

The definitions proposed for the model of innovation ecosystem are indeed varied. The model of an innovation ecosystem underlines the close relation and the interdependence between actors (Jackson, 2011). Some works assess the objectives of such an ecosystem, such as Gomes (2018), who underlines that the outcome of the ecosystem is about co-creation leading to the proposition of joint value. Tsujimoto (2018) details the form of the proposed value as either a product (e.g., cement produced from CE), a service (e.g., Shared waste collection), a process (e.g., the adaptation of an enterprise's processes to use a by-product), or organizational (e.g., the changes related to the internal enterprise to facilitate the exchange of resources). Some research focuses on the characteristics of an innovation ecosystem. Carayannis and Campbell (2009), for example, place more emphasis on the relationships between the different actors and more specifically the notions of collaboration and competition. Granstrand and Holgersson (2020), who proposed a systemic review around the model, put the accent on the three main components of an innovation ecosystem, actors, activities, and artefacts, which regroup tangible and intangible resources, which is a way to emphasize organizational innovation. The function of the actors is also analyzed, as in Guerrero et al. (2016), who discuss the innovation ecosystem and the key role of universities in the ecosystem development, or as in Nambisan and Baron (2013), who highlight the role of a pivotal actor (or key actor) in the innovation ecosystem. Several studies are also interested in the innovation ecosystem from a risk point of view linked to the integration of such an ecosystem (Adner, 2006), the viability of such an ecosystem (Walrave et al., 2018), or its stages of development (Dattée et al., 2018).

Based on these works, we can consider that an innovation ecosystem is a set of actors coming together around a common project, developing complex collaborative relationships that do not exclude competition (coopetition) and sharing resources for the creation of value which takes the form of various kinds of eco-innovation (product/service, process, organization) (see Table 1).

The innovation ecosystem is presented as an evolving system (Dattée et al., 2018). Throughout its functioning, actors will, for example, integrate or withdraw from the ecosystem, which will impact it. In this sense, an innovation ecosystem could have a life cycle that could resemble this evolution, consisting of three stages: initial stage, development stage, and renewal stage (Heaton et al., 2019) (Table 2).

The initial stage could be associated with the formation of the innovation ecosystem. The innovation ecosystem begins to take shape with an actor laying the foundations of the future ecosystem by defining a common project. Thanks to this first step, actors begin to join the ecosystem and generally revolve around the actor behind the initiative, who can later become the key actor (pivotal actor) of the

Table 1 Classification of the articles linking CE and ports

Topics	Examples	Examples between the 39 references
Construction of industrial symbiosis (IS)	– Waste from ships to be used in bio plants – Plastic waste for 3D printing – Fatal heat for district heating and shrimp farming – Sediment for cultivation – Contaminated dredging materials to create new lands	Karimpour et al. (2019) Garrido et al. (2021) Baas and Huisingh (2008) Martínez-Nicolás et al. (2021) Carpenter et al. (2018)
Urban metabolism	– Importance of cultural heritage toward CE – Studies of energy flows toward an energy transition of port cities – Implementation of smart urban cities in port territories – Effect of CE on urban regeneration	Acri et al. (2021), Fusco Girard (2013), Pintossi et al. (2021) Bahers et al. (2020), Bosman et al. (2018) Shahrokni et al. (2015) Williams (2019)
Preventive measures against pollution	– Results of energy and water reduction and the use of bio sludge – Reduction of emissions – Effects of CE on the environment – Management of solid waste from ports to reduce marine pollution	Baas (1998) Boehme et al. (2009) Budoc (2017) Uche-Soria and Rodríguez-Monroy (2019)
Innovation—new practices toward CE	– Proposition of new frameworks to help ports and cities to collaborate – Eco-designs for the rehabilitation of the old port quarry – Adoption of "green technological" innovations in the port – Innovative milieu and attractiveness	Roberts et al. (2021) Escobal et al. (2021) Lorek (2012) Kasmi (2020)

Sources: Authors, based on Veyssière et al. (2021)

ecosystem. The actors (mostly enterprises but also research institutions) remain few in number, and interactions between these same actors are quite rare at this stage, which can be explained by the time needed for the establishment of relationships but also that each actor must find its place in the ecosystem. The next stage that follows is ecosystem development; it is at this level that the ecosystem develops the fastest, interactions between actors become more intense, new actors join the ecosystem, and institutions integrate and assume a more active place in the ecosystem. Value is created both at the microlevel and the meso-level. The renewal stage is when actors find their places and functions within the innovation ecosystem. This stage is akin to the equilibrium point of the ecosystem. New actors are becoming rarer but continue to integrate into the ecosystem, and it is especially at this stage that the actors together begin to diversify their activities to co-evolve with the entire ecosystem (Table 3).

Our hypothesis is thus that the "circular" innovation ecosystem can take on the characteristics of an innovation ecosystem, namely, actors, relationships, resources,

Table 2 Characteristics of an innovation ecosystem

Actors	The actors in an innovation ecosystem can be divided into three groups: (1) enterprises and financial organizations, (2) research organizations, and (3) regulatory organizations. – Enterprises include start-ups to large enterprises, and financial organizations include commercial banks, crowdfunding platforms, etc. At the microlevel, the enterprises will therefore adapt their business models to match the common objectives of the innovation ecosystem (Konietzko et al., 2020; Lopez et al., 2019; Pieroni et al., 2019). – Research organizations, as they are public or private, are often seen as having leading research capabilities that are established in response to a particular need and can be viewed as important enablers of innovation (Philbin et al., 2014). The common role of these research institutions is the production of scientific knowledge and the diffusion of the knowledge generated to partner organizations. – Regulatory institutions or regulatory agencies are public organizations or government agencies that are set up to exercise a regulatory function. Their main role is to impose requirements or restrictions or to set the standard for activities. Regulatory institutions aim to create a business climate that will establish a favorable environment for the development of entrepreneurial activities (Uzunidis, 2014). In such an innovation ecosystem, among all the actors involved, a key actor (or pivotal actor) usually emerges and can be any actor within the innovation ecosystem, without being limited to a firm as for the business ecosystem. This pivotal actor exercises considerable influence over the strategies and fortunes of all the other actors (Nambisan & Baron, 2013; Nalebuff et al., 1996).
Relations	All the complementary actors within an innovation ecosystem are in close and interdependent relationships among themselves. By relations, we mean the interactions that are formed between the actors. These relations can take two forms: cooperation and competition. Actors are joining an innovation ecosystem to share common objectives and to gain benefits from it; to capture these benefits, they establish cooperation relationships that help facilitate deals when the need arises and create a source of opportunity (Jackson, 2011). The cooperation is both social and behavioral, including, for example, trust, openness, transparency, tact, professionalism, and complementarity. Cooperation relies on various forms of proximity among actors. The combination of different forms of proximities is necessary in order for cooperation between actors to be constituted. These forms of proximities can be geographical, organizational, cognitive, and institutional (Torre, 2009; Uzunidis, 2010). Although the actors are involved in cooperative relationships, these same actors still have different attributes, decision principles, beliefs (Tsujimoto et al., 2018), and own networks, with individual characteristics and objectives that also lead to competition. The cooperative relationship stems from the potential competitive advantage that an ecosystem can bring to these members. Synergies between actors allow access to new or more attractive resources (price, accessibility of the resource). However, cooperative relations do not prevent the existence of competition between those same actors. The relations of cooperation and competition are indeed present within an innovation ecosystem, and this is also known as coopetition. The

(continued)

Table 2 (continued)

	coopetition relationships between actors, which appears at first sight contradictory, allow the innovation ecosystem to evolve (Volkmer Martins et al., 2019).
Resources	To materialize the benefits from the relations, the actors need to commit resources toward the shared objectives (Autio & Thomas, 2014). By resources, we mean all the tangible and intangible assets used to produce goods, to provide services, and to facilitate exchanges or the entry of new actors. Resources can thus be divided into two categories: material (e.g., a product) and immaterial (e.g., a know-how) (Veyssière et al., 2021, Kasmi, 2018). In an innovation ecosystem, the resources can be (1) the workforce (students, staff, industry researchers, industry representatives, skilled workers) from the types of actors we mentioned earlier who will be in support of the new shared objectives (Heaton et al., 2019), (2) a financial form that is provided by financial organizations, and (3) know-how and competencies shared by actors within the ecosystem. But the resources can also take the form of inputs, wastes, or second raw materials and are valuable only because of the existence of the synergies. In this sense, synergies through the relations created within the innovation ecosystem can reveal resources that cannot be found elsewhere (Veyssière et al., 2021). Ideally, an innovation ecosystem based on a CE project will be structured to efficiently recover, reuse, and recycle any resources (including human capital) produced by all the actors. The way the resources are consumed (Jackson, 2011) and the continuous circulation of the flow (knowledge, energy, workforce, the raw material, etc.) within the innovation ecosystem determine its success or failure.
Results: Eco-innovations	Actors are joining an innovation ecosystem by establishing relations and committing resources together to gain benefits around the shared objectives. The benefits are then the results or outputs which take the form of value creation which cannot be created unless all the conditions (actors, relations, and resources) are present (Hannah & Eisenhardt, 2018). The creation of value in an innovation ecosystem adopting "circular" activities can be located at two levels, first at the microlevel for the enterprises of the ecosystem. This value creation generally takes the form of product eco-innovations (including services), processes, businesses, and organizations (Autio & Thomas, 2014; Gomes et al., 2018; OECD, 2018). The value can also be at the meso-level, that is to say all the actors of the territory in question. The value created can be tangible, like new resources (products, processes, etc.) created by the new projects or new shared infrastructures and circular jobs, but also intangible, for example, a trained workforce with knowledge and skills in the CE project, attractiveness for entrepreneurs, and start-ups (Nambisan & Baron, 2013). The creation of value can also be understood through the positive environmental impacts resulting from the CE project.

and results. The addition of CE in the model underlines the fact that the origin of this ecosystem is the CE project that connects actors around it and fuels its functioning. It brings together actors with the common goal of creating value at the micro- (enterprises) and meso- (ecosystem, territory) levels that takes the form of eco-innovations (products, services, organizations). The creation of such ecosystems

Table 3 Life cycle of an innovation ecosystem

Initial stage	Development stage	Renewal stage
• A key actor (pivotal actor) • Few enterprises • Few research institutions • Rare interactions between the actors of the ecosystem • Low identity	• Regulatory organizations and research organizations become active participants • Communication channels formed to exchange knowledge in the ecosystem • Inter-enterprises and professional networks become well established • Value creation at the micro- and meso-levels	• Stagnation of established sectors • Patenting activity shifts toward new domains

Source: Adapted from Heaton et al. (2019)

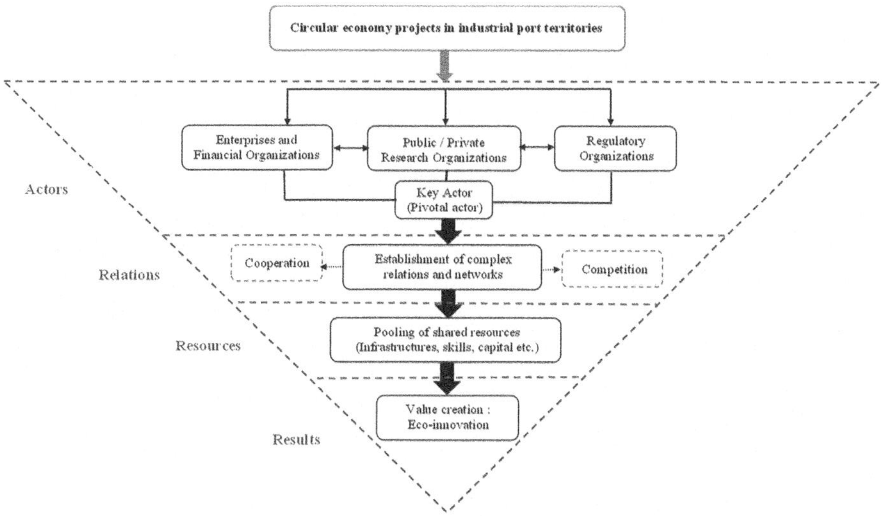

Fig. 2 Circular innovation ecosystem

requires that enterprises evolve their business models to align with the ecosystem's objectives. As such, companies' BM will be interconnected, according to the goals of the ecosystem (Shaw & Allen, 2018), oriented toward eco-innovation and more, largely sustainable, development. The following figure summarizes our analytical framework for industrial port territories which will be applied to the case of Dunkirk in France for this study (Fig. 2).

2 Methodology

2.1 Presentation of the Dunkirk Territory

Dunkirk is a coastal town located in northern France, on the border with Belgium and 100 kilometers from England. The city of Dunkirk is the fifth largest city in the Hauts-de-France region with 84,933 inhabitants in 2021. The Urban Community of Dunkirk (CUD) in which industrial ecology activities take place has no less than 195,917 inhabitants divided into 17 municipalities (Urban planning agency Flandre-Dunkirk, 2020).

The industrial port territory of Dunkirk, as it is today, began to form after the Second World War. The development of the steel industry in Dunkirk has its roots in the early 1960s, which was part of a vast national modernization program toward development, which also led to the creation of a second site of the same nature, in Fos-sur-Mer, Marseille. In Dunkirk, this program materialized with the installation in 1962 of the steel giant Usinor (now ArcelorMittal). For the plant's electricity needs, a thermal power station using Usinor's steel gases was installed. This marked the beginning of what would later be called industrial ecology. In the years that followed, enterprises gradually began to establish themselves in Usinor's slipstream. In 1985, industrial ecology practices extended beyond the private sector with the construction of the district heating network using the recovery of waste heat from Usinor's blast furnaces to supply the city. In 1990, the industrial environment plan integrating environmental concerns into industrial development was drawn up. In 1994, the CUD committed to a perspective of sustainable development and in particular industrial ecology. Since then, industrial synergies have been built in the industrial port territory of Dunkirk.

Industrial ecology in Dunkirk has evolved well since its inception, as shown in Fig. 3, so much so that in 2001, a collaborative project initiated by public and private actors in large industrial areas led to the creation of the Ecopal association, the first industrial ecology association in France. Industrial ecology in Dunkirk is based on flows of pooling and substitution.

The pooling flows on the territory mainly revolve around the collection of solid industrial waste such as reel offcuts and scrap metal (Ball Packaging), waste of the paper type (soiled or not) (Aluminum Dunkirk, Chaudronnerie ADS, DK6, Ryssen Alcohol, Dillinger), or liquid industrial waste such as oils and paint waste (Port of Dunkirk, Dillinger). Substitution flows revolve essentially around gas recovery (ArcelorMittal, Baudelet environment, DK6, Ryssen Alcohol), slag recovery (ArcelorMittal, Befesa Valera), fatal heat (ArcelorMittal, Dalkia), or even sediments, water, and industrial sludge (Suez Eau France, EDF, Daudruy, port of Dunkirk, DK6).

To verify the existence of a circular innovation ecosystem in the industrial port territory of Dunkirk, data collection was necessary in addition to the data already available.

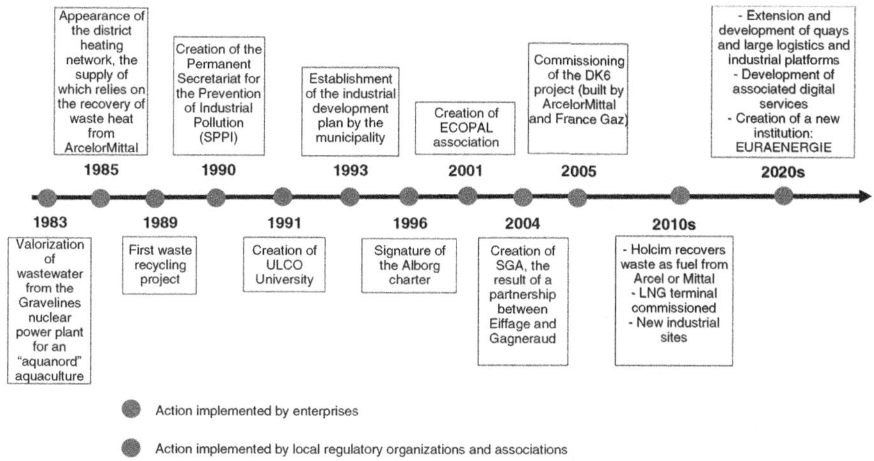

Source: Adapted from Boutillier et al. (2015) and (Kasmi, 2018)

Fig. 3 The main lines of the history of industrial ecology in Dunkirk. Source: Adapted from Boutillier et al. (2015) and Kasmi (2018)

2.2 *Data Collection*

The data used in this research is a combination of primary and secondary qualitative data that allow us to study the industrial port territory of Dunkirk and the "circular" innovation ecosystem that could be found there.

The collection of primary data is based on the IMPPEC research program[3] which aims to create a table of indicators to highlight and assess the role of the port of Dunkirk in the industrial ecology and the circular economy. The specificity of these indicators is that they combine traditional indicators associated with industrial ecology/circular economy (economic, social, and environmental impacts) but also more specific indicators related to the link between port activities and the circular economy.

The indicators of the IMPPEC program focus on four axes:

Axis 1—Strategy, governance, communication evaluating the operational strategy of the port in support of circular economic development

Axis 2—Industrial symbiosis measuring the development of industrial symbiosis at the level of the territory

[3] Funded by the large seaport of Dunkirk (GPMD) in collaboration with the Urban Community of Dunkirk (CUD) and the institution ECOPAL (partner economy and ecology in local action). This program is carried out by the "Innovation and Industrial Strategies" (ISI) laboratory of the Université du Littoral Côte d'Opale (ULCO).

Table 4 List of interviews

Type of actor	Targets numbers	Responses received	Participation rate (%)	Interviewees
Enterprises	66	32	48.49	Representatives of the enterprise
Public/private research organizations	16	8	50	University training managers at ULCO[a]
Regulatory organizations	16	7	43.75	Representatives of the organization

[a]University of Littoral Côte d'Opale (ULCO) with 4 teaching sites and over 300 researchers and 10,000 students. ULCO was targeted for our research in which we conducted surveys among the training managers (pre-selected) of each department. The university is present on four sites: Dunkirk, Calais, Boulogne, and Saint Omer

Axis 3—Economic development measuring the means or resources invested/developed for the development of the circular economy and the impacts in economic terms of practices relating to the circular economy

Axis 4—Environmental impacts assessing the reduction of negative externalities caused by activities from industrial port territories

Collecting primary data required carrying out surveys using dedicated questionnaires for each type of actor that was previously selected using data already available in the territory; a flexible approach had to be used for data collection. The questionnaires were carried out using LimeSurvey software and a Google form and then sent to the actors concerned. Personalized follow-up was then provided to allow the actors to complete the questionnaires (telephone call and videoconferences to provide details and face-to-face meetings when possible).

To verify the characteristics of a "circular" innovation ecosystem, preliminary sorting work was carried out using the data available with the help of the CUD, Ecopal, and the port of Dunkirk to establish the list of target actors. The following table presents the list of target actors interviewed (Table 4).

The results of the questionnaires were then analyzed with the same software, not only to verify the presence of a circular innovation ecosystem through the different characteristics highlighted (namely, actors, relationships, pooled resources, and results) but also to situate the stage in which the innovation ecosystem in Dunkirk is at (initial stage, development stage, or renewal stage) while highlighting the limits and difficulties accompanying the transformations of the industrial ecology and highlighting the opportunities that accompany these transformations (Table 5).

Table 5 Selected indicators to verify the existence of a circular innovation ecosystem

Characteristics of a "circular" innovation ecosystem		Axes of the IMPPEC program concerned	Selected indicators
Actors	Enterprises and financial organizations	Industrial symbiosis	Number of enterprises in the IS* Key actor
	Regulatory organizations	Strategy, governance, and communication	Initiatives created for the CE Form of governance Communication around CE and sustainable development
	Public/private research organizations	Industrial symbiosis	Number of training courses* Number of laboratories*
Relations	Enterprises and financial organizations	Industrial symbiosis	Number of synergies between enterprises*
	Regulatory organizations		Involvement in synergies between enterprises
	Public/private research organizations		
Resources		Economic development (means)	Land available Infrastructures made available R&D expenditure
Results at the micro- and meso-level		Economic development (Results)	Eco-innovations Circular job created Number of start-ups
		Environmental impacts	Resource saving Reduction of emissions*

3 Findings and Discussion

3.1 Findings

3.1.1 Actors

The IE project in Dunkirk began in 1985 with the district heating project supplied by the fatal heat of the enterprise ArcelorMittal. Cornerstone of the IE in the territory, this project has opened the doors to exchanges of flows between enterprises which continue to increase in density. If IE started with ArcelorMittal, this enterprise has over time become the key actor within the IS due to its strategic position as the largest supplier of co-products such as slag or fatal heat (e.g., 12 enterprises reuse its co-products directly in the IS). Currently, the IS in Dunkirk is made up of no less than 44 enterprises of varying sizes and is mainly grouped around the activities of metallurgy, iron and steel, chemicals, construction materials, and the aqua business.

These enterprises in the IS are the main employer in the territory of Dunkirk, with, for example, ArcelorMittal (3641 employees), EDF (1830 employees), Aluminum Dunkerque (570 employees), Dillinger France (550 employees), Harsco Metal (471 employees), Ascometal (304 employees), and Minakem (230 employees).

The other enterprises present are mainly subcontractors with a workforce of less than 80 employees (AGUR, 2019). By adopting CE activities, these large enterprises are developing their business models so as to be able to accommodate circular activities. This evolution of the business model into a business model innovation based on circular economy practices includes the adoption of a co-product in its process to produce eco-innovation, the sharing of resources, but also of know-how.

The IE in the territory, and, more widely, CE, is supported by financial organizations such as the European Union, the French State, and the regional council, which have decided to allocate funds to support the revival of the economy and the de-carbonization of businesses in the territories, including Dunkirk. Thus the European Union launched the "just transition fund[4]" amounting to 40 billion euros for the European regions most dependent on fossil fuels, while the French State, with "France Relance[5]" (from 2020 to 2022), allocates an envelope of 100 billion euros dedicated to ecological transition. More locally, the Hauts-de-France region offers an additional fund endowed with 1.3 billion euros for the sustainable economy (Dunkerque Promotion, 2021). Dunkirk is one of the 24 winners of the call for submissions for the "Territories of Innovation and Great Ambition" (TIGA) action and the only winner from the Hauts-de-France region (2018). With its project entitled "Transformation of an industrial port ecosystem," Dunkirk responds to the issue of dependence on carbon energy and therefore benefits from a portion of the 450 million euros spread over 10 years for the selected territories of national interest.

The CE is supported by regulatory organizations which set up a favorable climate for the development of circular activities in the territory. Desiring the transition to happen, these regulatory organizations have also turned to sustainable development and are adapting their objectives to support the transition to the CE. Among the organizations questioned, five are directly developing initiatives for local and IS enterprises for the transition to the CE, taking the form of support for enterprises to find premises and land and to provide information in the form of industrial map, advice, and follow-up. Territorial governance is present and begins to coordinate around CE projects. This coordination is dynamic (which means changing, collective, and evolving) and leads to communication actions through the available media such as the websites of organizations, local newspapers, social networks, or even information sessions and conferences held regularly in the territory (Kasmi et al., 2017). The representatives of the port of Dunkirk, for example, affirm in its development strategy that it wishes to turn to the CE from 2009 with its planning and sustainable development plan (PA2D). At the university level, the IMPPEC program studies the impact of ports on the CE, the objectives being to anticipate and plan the challenges for the future for Dunkirk Port, both for economic development objectives and for objectives of the sustainable management of the territory (initiated at

---

[4]The Just Transition Fund supports the economic diversification and reconversion of the territories concerned. https://ec.europa.eu/regional_policy/en/

[5]https://www.dunkerquepromotion.org/soutenons-les-entreprises-locales/plan-de-relance-national-france/

Table 6 Findings for the actors

Characteristics of a "circular" innovation ecosystem		Selected indicators	Findings
Actors	Enterprises and financial organizations	• Number of enterprises in the IS* • Focal actor	• 44 enterprises • An enterprise: ArcelorMittal
	Regulatory organizations	• Initiatives created for the CE • Form of governance • Communication around CE and sustainable development	• In the form of follow-up and information provided by five organizations • Dynamic form • Multiple channels like newspapers, conferences
	Public/private research organizations	• Number of training courses* • Number of laboratories*	• Four Master's courses in connection with CE • 14 laboratories within ULCO

the national level through the measures taken, territorialization of activities, reduction of pollutant emissions, particularly atmospheric, reduction of waste production, and job creation).

The research organization in the territory of Dunkirk is the Littoral University (ULCO). With 38 Master's courses, the university has adapted to the territory's transition by offering 4 Master's courses in connection with the CE with an average of 69 students per year in Master's 2. Nevertheless, the university, in addition to its role of trainer, is above all a pool of 300 research professors who are spread over 14 laboratories and technological scientific platforms around three research poles: (1) sea and coastline, transformations, and challenges, (2) technological and environmental mutations, and (3) integrated humanities and territories. Despite the presence of these laboratories, collaborative relationships with other actors in the territory remain marginal to date. Research requiring researchers is generally conducted outside the territory and then applied there, which can be explained by the presence of production units of multinational corporations whose decision centers are located in other cities or countries (Laperche et al., 2021). An example in the field of industrial ecology is the EPIFLEX solution developed by EDF[6] R&D and Mines ParisTech,[7] which proposes an innovative approach allowing the design of industrial eco-parks that enhance the flow of materials and energy.

These findings are summarized in Table 6.

[6]Électricité de France (EDF) is a French enterprise, producing and supplying electricity, more than 80% owned by the State. The enterprise is the leading producer and the leading supplier of electricity in France and Europe.

[7]"École nationale supérieure des mines de Paris," also known under the names of "École des mines de Paris" or simply "les Mines" and since 2008 MINES ParisTech, is one of the 204 French engineering schools accredited as of September 1, 2020, to deliver an engineering degree.

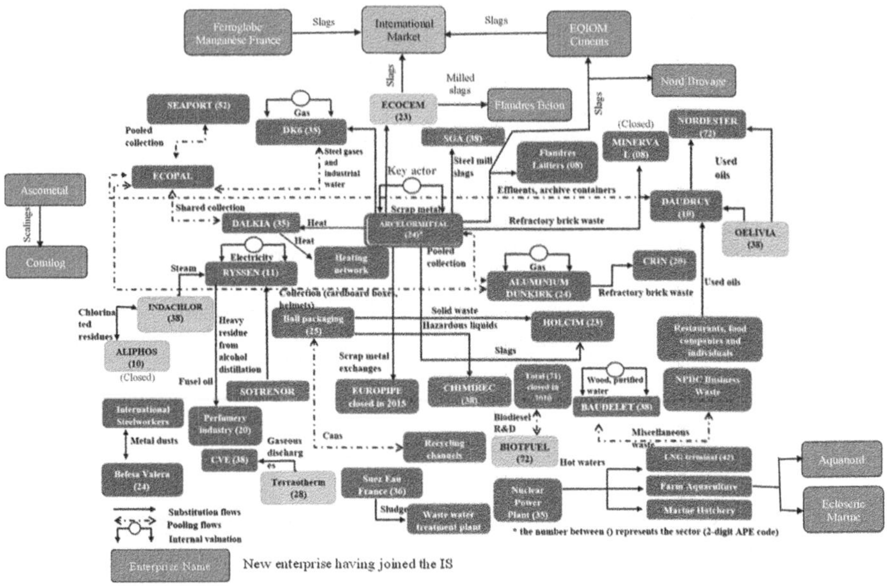

Source: Based on Kasmi (2020) updated by the authors.

Fig. 4 Industrial ecology in the industrial-port territory of Dunkirk. Source: Based on Kasmi (2020) updated by the authors

3.1.2 Relations

The enterprises in the IS of Dunkirk establish collaborative relationships that take the form of synergies through the exchange of substitution and pooling flows. In Fig. 4, we highlight these relationships between enterprises through the flows that circulate among them. For the pooling flows, industrial water, gas, solid, and miscellaneous waste are the most common flows that circulate in the IS. For example, ArcelorMittal and Aluminum Dunkerque are undertaking pooled collection, while Baudelet and NPDC Business Waste deal with miscellaneous waste. Substitution flows are the most common flows in the IS of Dunkirk. With district heating using fatal heat from ArcelorMittal which marks the foundation of IS, this fatal heat from ArcelorMittal is now also used by other enterprises such as Dalkia and its steel gases by DK6. ArcelorMittal's slag is reused by several enterprises including ECOCEM, SGA, Flandre Laitier, and EQIOM Cement, for example. The other substitution flows that are often found in the IS along with the slags or fatal heat are mostly used oils and chemical residues.

These synergies are facilitated by the regulatory organizations. These regulatory organizations are involved in each stage of the development of synergies between enterprises in the CE. These stages are emergence, development, testing, actual implementation, and sustainability. Regarding the survey conducted, all the

Table 7 Regulatory organizations and their interventions in the development of synergies

Regulatory organizations	Stage of development of synergies				
	Emergence	Development	Testing	Implementation	Sustainability
Port of Dunkirk	✓	X	X	X	X
Dunkerque Promotion	✓	✓	✓	✓	✓
Initiative Flandre	✓	✓	X	X	X
Dunkirk Water Syndicate	✓	✓	X	✓	✓
AGUR	✓	✓	✓	X	X
ECOPAL	✓	✓	✓	✓	✓
Euraénergie	✓	✓	✓	✓	✓

regulatory organizations questioned are involved in synergies between enterprises across the CE, in at least one of the five stages mentioned above (Table 7).

The ECOPAL association, for example, is involved in all stages of CE activities in Dunkirk and plays a central role in the collection of waste within IS. The regulatory organizations also develop collaborative relations among themselves to propose a better solution for CE implementation. "Euraénergie" is one these collaborations, organized around five administrators: the CUD, the Urban Planning Agency (AGUR), Dunkerque Promotion, Pôlénergie, and ECOPAL. This tool unites public and private actors around common objectives and offers a range of services allowing the promotion of the establishment of innovative enterprises, to attract and support project leaders while supporting research and innovation to promote transitions in the industrial port territory of Dunkirk. The port also has relations with local regulatory organizations such as the AGUR around exchanges and sharing information relating to the different specific maps of the territory (industrial, water, energy, agricultural, etc.) and ECOPAL around exchanges and sharing information relating to various local subjects on the CE and waste management. Within the Hauts-de-France region, the port supports the development of projects linked to CE including eco-materials and CE[8] and ULCO around research projects (e.g., IMPPEC Program).

Figure 4 is a representation of IS in Dunkirk based on the work of Kasmi (2020). Gray boxes are enterprises in IS found in previous work, while blue boxes are enterprises newly integrated in IS, obtained thanks to the results of surveys. The details regarding Fig. 4 are further developed in the following sections.

3.1.3 Resources

The actors who establish relationships with each other have resources available which make it possible to welcome new synergies or new businesses. Land reserves

[8] Hauts-de-France region (2021), https://www.hautsdefrance.fr/economie-circulaire/

Table 8 Findings for the resources

Characteristics of a "circular" innovation ecosystem	Selected indicators	Findings
Resources	• Land available* • Infrastructure made available • R&D expenditure	• 3000 hectares • Innovation park, "Hôtel des technologies," Digital lab, and facilities • From 1% to 20% distributed among eight enterprises

are part of these resources. The territory of Dunkirk has a substantial land reserve for the development of activities. With 3000 hectares of available land, the territory has 500 hectares of immediately available land and 125 hectares of "plug-and-play" land in the large industrial zone, of which 45 hectares have already been reserved (Dunkerque Promotion, 2021).

In addition to land, infrastructure is available: a 5-hectare innovation park operational by the end of 2022 dedicated to energy and CE transition. This park is meant to allow project leaders to test their innovations with one-to-one scale prototypes and also to be in contact with ULCO research laboratories and two engineering schools (Engineering School of Littoral Côte d'Opale (EILCO) and Institute Mines Telecom (IMT Nord Europe)). The newly constructed "Hôtel des technologies" located in central Dunkirk hosts ArcelorMittal's Digital Lab and is made up of pre-equipped cells available for potential start-ups in industrial and energy innovation. In addition, the Grand Port Maritime of Dunkirk provides its facilities to accommodate all types of goods and the largest ships. With a rail network, a network providing access to the European motorway network, and a large gauge canal, Dunkirk is to date the leading French rail freight hub and the leading river port in the Hauts-de-France region (Dunkerque Promotion, 2021).

However, among the 44 enterprises that responded to our surveys, only 8 enterprises spend between 1% (7 enterprises) and 20% (1 enterprise) on R&D (the largest enterprises being those that spend the least on R&D). We note that Dunkirk is a territory for implementing innovations that are developed elsewhere and that enterprises invest only a little or not at all in R&D (Table 8).

3.1.4 Results: Eco-innovations

The output of these collaborative relations takes the form of eco-innovation that can be found at the micro- and meso-level, however in a weak proportion. At the microlevel, CE projects have enabled enterprises to create synergies by changing their BMs. The eco-innovations can take the form of new or improved goods. Out of the 44 enterprises questioned, only 10 enterprises introduced new or improved products, for example, slags and co-products like scaling allow enterprises to produce CE products. ArcelorMittal's slags, for example, enable enterprises like Nord Broyage, Eqiom cements, SGA, or ECOCEM to manufacture low carbon cement. ECOCEM produces "green" cement from this slag, which releases 45 times less CO_2 than traditional cement during the production.

Other eco-innovations at this level can be manifested as a change in the production process. Only 10 enterprises had made this type of eco-innovation; these enterprises have adapted and transformed their production processes to be able to reuse the co-products. Indachlor, for example, has revised its processes by creating a short circuit waste recycling unit to transform liquid waste using an incinerator to produce high-quality steam which will then be reused by IS enterprises.

Another expected form of eco-innovation is new or improved services and new forms of work organization. Only five enterprises answered positively, which represent 15% of the responses obtained. Although our expectations for the forms of eco-innovation also include new commercial methods, only one enterprise (3%) is currently processing the changes to propose new commercial methods.

At the meso-level, some eco-innovations are also worth noticing. The eco-innovation here takes the shape of collaborative tools. For example, the "Recycle solution[9]" created by ECOPAL is a digital tool for managing waste and recoverable products adapted to all sizes of companies. The SHYMED project resulting from the collaboration between two enterprises (Hynamics and EDF) provides for the creation of a green hydrogen production and distribution station to supply the buses and dumpsters in Dunkirk. The project is expected to be in service by January 2024.

The "Little Cithy[10]" future project piloted by ENGIE in partnership with the CUD is a project at the stage of feasibility with, for the moment, no commissioning date. The project is supposed to provide for the use of green urban gas with a significant reduction in gas emissions of 6% for CO_2 (20 tons CO_2eq. avoided per year) and up to 63% for CO.

The adoption of circular activities is also bringing changes, leading to the creation of jobs within the IS that also create value at the scale of the territory. Among the enterprises surveyed, 14 enterprises have created jobs over the past 5 years (890 jobs created). Nevertheless, out of these 890 jobs created, only 44 jobs are directly related to the CE, which represents barely 5% of new jobs in the IS. This creation of jobs is not the fact of start-ups, which still remain inexistent, but of the large enterprises already present on the territory.

The results obtained from CE activities are also reflected in environmental impacts. By taking into account only the following synergies, heat network, Ecocem, shared collections, LNG terminal, and nuclear power plant, the industrial symbiosis allowed the recovery, on average, of 5 billion m^3 of steel gas, 994,900 tons of CO_2 per year avoided, 150.4 GWh of waste heat recovered (16,000 Equivalent housing), and 750,232 tons of materials recovered (Table 9).

In terms of gas emissions reduction, most of the enterprises produce less.

---

[9] https://recycle-solution.fr/

[10] https://www.dunkerquepromotion.org/2021/10/vers-le-deploiement-dune-filiere-hydrogene-a-dunkerque/

Table 9 Findings for the eco-innovation

Characteristics of a "circular" innovation ecosystem	Selected indicators	Findings
Results at the micro- at meso-level	• Eco-innovations: – New or improved goods – New or improved production process – New or improved services and new forms of work organization – New commercial methods – Collaborative tools • Circular job created* • Number of start-ups	• 10 enterprises out of 44 • 10 enterprises out of 44 • 5 enterprises out of 44 • 1 enterprise out of 44 • 44 jobs • 0 start-ups
	• Resource saving* • Reduction of emissions*	• Energy and raw materials • Detailed in Table 10

Table 10 Reduction of emissions per enterprise based on surveys during the last 5 years

Enterprise	Emission reduction
EQIOM	5% on gas emissions
Bois Environnement Services	25% on gas emissions
NORTH ESTER	10% on CO_2
IMERYS Aluminates	2.5% on CO_2
Terraotherm	30% on CO_2
COMILOG	CO_2: 5% and dust: 5%
DAUDRUY	20% on CO_2
ECOCEM	680 kg less of CO_2 per ton of cement produced
GRDF	The use of biomethane divides CO_2 emissions by 10 compared to fossil gas
Dunkerque LNG and Gaz-Opale	100,000 tons per year less (average over the last 5 years)
ArcelorMittal	SO_2—Emissions of sulfur dioxide (kg per ton of liquid steel) are decreasing and amounted to 0.93 kg per ton of steel in 2017 against 1.26 in 2015 NOx—Emissions of nitrogen oxides (kg per ton of liquid steel): 1.29 in 2017 against 1.33 in 2015

To go further, the changes made so far are currently inspiring new projects that continue to transform the territory. The collective territorial project "heat highway," piloted by Pôlénergie[11] with Euraénergie,[12] is undoubtedly the most ambitious project in terms of CE in the territory. This large-scale project aims to recover the fatal heat from energy-intensive sites in the port of Dunkirk to distribute it to other enterprises. The heat highway will be added as an extension to the district heating network already present in the territory. This project is the winner of the Responsible Sourcing award of the 2021 Sustainable Industry Trophies. Moreover, Dunkirk is now welcoming new projects like the Gigafactory for the development of low-carbon battery cells by the enterprise Verkor. This company will create many jobs in the territory, but we can mention that R&D activities will once again not be located in Dunkerque.

3.2 Discussion

The characteristics of a circular innovation ecosystem highlighted through the literature, namely, the actors, relationships, resources, and results, are found in the industrial port territory of Dunkirk. The circular economy project in the territory took shape with the enterprise ArcelorMittal (ex-Usinor) and the first synergy developed, the reuse of fatal heat for the urban heating network. This CE project has become more structured with the integration of new enterprises into the IS but also with an increasingly active involvement of regulatory organizations to facilitate the development of new synergies and at the same time create a favorable climate to attract new enterprises, while research organizations provide their know-how, expertise, and training courses. All these actors work and co-evolve with the ecosystem. The adoption of circular activities has also changed the way enterprises in IS create value. By having access to pooling and substitutions flows and by integrating by-products into its value creation chain, enterprises welcome new possibilities, which has the effect of reorienting their business models toward a more sustainable business model. This change in business models at the individual level of each enterprise aims to achieve a common goal, centered on a circular project taking the form of eco-innovations.

However, in the case of the industrial port territory of Dunkirk, even if all the ingredients seem to be in place to emerge eco-innovations—actors (enterprises that integrate and adapt their business models, regulatory organizations that support and promote, and research organizations helping on the research side), intensified relationships between these actors, and shared resources—the results do not seem to match. Technological eco-innovations remain negligible compared to the efforts put

[11] Created in 2011 as a result of the desire of the Regional Council of Nord-Pas de Calais and the Urban Community of Dunkirk, Pôlénergie wants to translate the Energy Transition into an economic opportunity for businesses and territories in Hauts-de-France.

[12] Euraénergie supports the transformation of the Dunkirk territory within the framework of the project "Dunkerque, l'Énergie Créative" while contributing to the energy strategy through its innovation park.

in place. In terms of job creation, although the enterprises in IS have created several jobs over the past 5 years, the share of circular jobs remains low (5% of the jobs created within the IS). Some circular economy projects set up in Dunkirk were developed in partnership with universities outside the territory, which shows a difficult integration of the research capacity of ULCO, which is hindered on projects that are established in its territory. It should also be noted that, despite the adoption of circular activities and the integration of enterprises into IS, the closure of some enterprises introduces uncertainties into the solutions provided by the ecosystem (like competitive advantage). We also note that the enterprises forming IS are generally large enterprises; the absence of start-ups in the ecosystem raises questions about the innovation dynamics fostered by this ecosystem. Governance in the territory, which has just passed from a static form to a more dynamic form, is still struggling to coordinate the efforts of actors optimally and regulatory organizations continue to intervene at the same stages in the development of synergies.

The circular innovation ecosystem in Dunkirk, based on the actions implemented and the limitations put forward, just entered the development stage (according to Table 2) and is mainly characterized by regulatory organizations and research institutions taking a more active role within the ecosystem. Even if this active role is still in its early stages, a notable evolution is still observed. Regulatory organizations, with the collaboration of the other actors, multiply the communication channels for exchanging and sharing knowledge (setting up conferences, working groups between actors, channels to facilitate access to data, etc.). The ecosystem identity and relations between enterprises are well established (around the by-products of steel and metallurgical activities). To continue to develop, the ecosystem could start by optimizing the role of each regulatory organization in the ecosystem. This action could make it possible to redirect efforts toward the stages of the creation of synergies for which support is not yet developed. Giving a more important role to ULCO University so that the co-evolution of ecosystem actors takes place in a more homogeneous and balanced way is also an important step toward the building, and efficient functioning, of the circular ecosystem of innovation. In this vein, favoring the creation of start-ups (with the common support of large companies, the university, and institutional actors) could be a way to increase the innovative outcome of the ecosystem.

4 Conclusion

In this chapter, we propose to study the impacts of CE in industrial port territories. CE can become a new economic model for industrial port territories which face the challenges of complying with environmental constraints associated with the scarcity of resources and the increasing price of energy but also changes related to regulations. From the literature, we made the link between CE and the innovation ecosystem. Our findings show that CE can enhance and strengthen the characteristics of an innovation ecosystem.

Our work led to a definition of a circular innovation ecosystem as an innovation ecosystem fueled by a circular economic project. It is formed by a set of enterprises, financial organizations, public or private research institutions, and regulatory organizations that establish collaborative relationships and share resources related to the development of CE (shared infrastructures, know-how, and human resources). The enterprises involved in such an ecosystem transform and adapt their business model to be able to produce eco-innovations with the other actors of the ecosystem. These eco-innovations can therefore take the shape of product or service innovations, process innovations, or even organizational innovations. In such innovation ecosystems, there is the creation of value both at the micro- and meso-level, oriented toward circular economic activities.

From the design of the circular innovation ecosystem determined from the literature, we were interested in the case of the industrial port territory of Dunkirk and more precisely IS, which finds its origins in the 1980s, our objective being to determine if and in which ways a CE project could initiate a circular innovation ecosystem. Our results show that the characteristics of such an ecosystem are found in Dunkirk. There are indeed actors who work in collaboration and share resources to develop circular activities. Nevertheless, we notice that the eco-innovations produced in the territory do not translate to established coordination between the actors. Although the actors have the means to develop and carry out eco-innovations, the industrial port territory of Dunkirk remains a territory for the implementation of eco-innovations that are developed outside its borders. Regulatory organizations too often intervene at the same stages in the development of synergies, which reduces the impact of their actions and can even create confusion as to the role of each of them. The university seems to take a secondary role despite its proficiency and the formations set in place. A study on the optimization of the role of each institutional actor can therefore be considered for future research to deepen the reasoning behind why there is such gap between the efforts and results displayed. Another avenue to be explored would be to undertake case studies about the way companies transform their business models and identify the barriers in their transition to green business models.

Our work contributes to the development of a methodology for the assessment of the development and the evolution of circular innovation ecosystems. The next step will be to make comparisons between different ports to refine this methodology and highlight good practices in CE.

References

Acri, M., Dobričić, S., & Debevec, M. (2021). Regenerating the historic urban landscape through circular bottom-up actions: The urban seeding process in Rijeka. *Sustainability, 13*(8), 4497.

Adner, R. (2006). Match your innovation strategy to your innovation ecosystem. *Harvard Business Review, 84*(4), 98.

Adner, R., & Kapoor, R. (2010). Value creation in innovation ecosystems: How the structure of technological interdependence affects firm performance in new technology generations. *Strategic Management Journal, 31*(3), 306–333.

AGUR. (2019). *L'éconographie Flandre-Dunkerque.*

Andriamanantena, A., Veyssière, S., Le, S. T. K., Cotonnec, G., & Laperche, B. (2020). Le Rôle des Ports dans L'Économie Circulaire. Construction d'un Tableau de Bord d'Indicateurs. https://riifr.univ-littoral.fr/wp-content/uploads/2012/12/doc-312-Le-role-des-ports-dans-leconomie-circulaire.-Construction-dun-tableau-de-bord-dindicateurs.pdf

Autio, E., & Thomas, L. (2014). Innovation ecosystems. In *Oxford handbook of innovation management* (pp. 204–288).

Baas, L. (1998). Cleaner production and industrial ecosystems, a Dutch experience. *Journal of Cleaner Production, 6*(3–4), 189–197.

Baas, L. (2000). Developing an industrial ecosystem in Rotterdam: Learning by. . . what? *Journal of Industrial Ecology, 4*(2), 4–6.

Baas, L. W., & Huisingh, D. (2008). The synergistic role of embeddedness and capabilities in industrial symbiosis: Illustration based upon 12 years of experiences in the Rotterdam harbour and industry complex. *Progress in Industrial Ecology, An International Journal, 5*(5–6), 399–421.

Bahers, J.-B., Tanguy, A., & Pincetl, S. (2020). Metabolic relationships between cities and hinterland: A political-industrial ecology of energy metabolism of Saint-Nazaire metropolitan and port area (France). *Ecological Economics, 167*, 106447.

Blomsma, F., & Brennan, G. (2017). The emergence of circular economy. *Journal of Industrial Ecology, 21*.

Boehme, S. E., Panero, M. A., Muñoz, G. R., Powers, C. W., & Valle, S. N. (2009). Collaborative problem solving using an industrial ecology approach: The New York/New Jersey Harbor economy-wide substance flow case studies. *Journal of Industrial Ecology, 13*(5), 811–829.

Bosman, R., Loorbach, D., Rotmans, J., & Van Raak, R. (2018). Carbon lock-out: Leading the fossil port of Rotterdam into transition. *Sustainability, 10*(7), 2558.

Boutillier, S., Burmeister, A., Laperche, B., Brogniart, C. M., Uzunidis, D., & Kasmi, F. (2015). *Le territoire entrepreneurial durable.* Etude du cas de Dunkerque (Nord-France).

Budoc, R. L. (2017). Le développement des ports ultramarins: quels enjeux environnementaux? *Développement durable et territoires. Économie, géographie, politique, droit, sociologie, 8*(1).

Carayannis, E. G., & Campbell, D. F. (2009). 'Mode 3' and 'Quadruple Helix': Toward a 21st century fractal innovation ecosystem. *International Journal of Technology Management, 46*(3–4), 201–234.

Carpenter, A., Lozano, R., Sammalisto, K., & Astner, L. (2018). Securing a port's future through circular economy: Experiences from the Port of Gävle in contributing to sustainability. *Marine Pollution Bulletin, 128*, 539–547.

Commission, E. (2020). *Communication de la commission au parlement Européen, au conseil, au comité économique et social Européen et au comité des régions*. Bruxelles.

Dattée, B., Alexy, O., & Autio, E. (2018). Maneuvering in poor visibility: How firms play the ecosystem game when uncertainty is high. *Academy of Management Journal, 61*(2), 466–498.

Escobal, I., Manuel Baraibar, J., Garcia Mochales, J. L., Gonzalez, J., Arlandi, M., & Bernardo, A. (2021). Stabilization works in the quarry of Punta Lucero (Port of Bilbao). *Informes de la Construccion, 73*(562).

Foulquier, E. (2012). Politique (s) de l'espace portuaire. L'Espace Politique. *Revue en ligne de géographie politique et de géopolitique* (16).

Foundation, E. M. (2015). *What is circular economy?* https://www.ellenmacarthurfoundation.org/circular-economy/concept/schools-of-thought

Freeman, C. (1982). Technological infrastructure and international competitiveness', draft paper submitted to the OECD ad hoc group on science, technology and competitiveness. *Mimeo.*

Freeman, C. (1987). *Technology policy and economic performance lessons from Japan*. Frances Printer.

Frosch, R. A., & Gallopoulos, N. (1989). Strategies for manufacturing. *Scientific American, 261*, 144–152.

Fusco Girard, L. (2013). Toward a smart sustainable development of port cities/areas: The role of the "historic urban landscape" approach. *Sustainability, 5*(10), 4329–4348.

Gallaud, D., & Laperche, B. (2016). Économie circulaire et développement durable. Écologie industrielle et circuits courts. *Revue Projet*, 6.

Garrido, J., Silva, D., Portela, B., & Lekube, B. (2021). Integrated user-oriented service for 3D printing environments with recycled material from maritime plastic waste. *Applied Sciences, 11*(9), 3787.

Geldron, A. (2013). *Fiche technique-Économie Circulaire: Notions, version modifiée Octobre 2014*. Ademe.

Gomes, L. A., Facin, A. L. F., Salerno, M. S., & Ikenami, R. K. (2018). Unpacking the innovation ecosystem construct: Evolution, gaps and trends. *Technological Forecasting and Social Change, 136*, 30–48.

Granstrand, O., & Holgersson, M. (2020). Innovation ecosystems: A conceptual review and a new definition. *Technovation, 90*, 102098.

Guerrero, M., Urbano, D., Fayolle, A., Klofsten, M., & Mian, S. (2016). Entrepreneurial universities: Emerging models in the new social and economic landscape. *Small Business Economics, 47*(3), 551–563.

Guillard, V. (2018). D'une économie linéaire à une économie circulaire. *Economie et Management, 168*.

Hannah, D. P., & Eisenhardt, K. M. (2018). How firms navigate cooperation and competition in nascent ecosystems. *Strategic Management Journal, 39*(12), 3163–3192.

Heaton, S., Siegel, D. S., & Teece, D. J. (2019). Universities and innovation ecosystems: A dynamic capabilities perspective. *Industrial and Corporate Change, 28*(4), 921–939.

Homrich, A. S., Galvao, G., Abadia, L. G., & Carvalho, M. M. (2018). The circular economy umbrella: Trends and gaps on integrating pathways. *Journal of Cleaner Production, 175*, 525–543.

Jackson, B. (2011). What is an innovation ecosystem?.

Karimpour, R., Ballini, F., & Ölcer, A. I. (2019). Circular economy approach to facilitate the transition of the port cities into self-sustainable energy ports—A case study in Copenhagen-Malmö Port (CMP). *WMU Journal of Maritime Affairs, 18*(2), 225–247.

Kasmi, F. D. (2018). *Écologie industrielle, milieu éco-innovateur et diversification de l'économie territoriale: le cas du complexe industrialo-portuaire de Dunkerque*. Université du Littoral Côte d'Opale. Thesis. https://tel.archives-ouvertes.fr/tel-01998218/document

Kasmi, F. (2020). Industrial symbiosis and territorial development: The cross-fertilization of proximity dynamics and the role of information and knowledge flows. *Journal of the Knowledge Economy*, 1–21.

Kasmi, F., Laperche, B., Brogniart, C. M., & Burmeister, A. (2017). Écologie industrielle, milieu innovateur et gouvernance terri-toriale: Le cas de Dunkerque (Nord- France). *Canadian Journal of Regional Science*.

Konietzko, J., Bocken, N., & Hultink, E. (2020). Circular ecosystem innovation: An initial set of principles. *Journal of Cleaner Production, 253*, 119942.

Lacy, P., Long, J., & Spindler, W. (2020). *The circular economy handbook*. Springer.

Laperche, B., Fourmentin, S., Llorens, J. M., Hardouin, P., & Abi-Aad, E. (2021). Une approche systémique de la valorisation de la recherche : le cas de l'université du Littoral Côte d'Opale. *Technologie et innovation, 6*. https://doi.org/10.21494/ISTE.OP.2021

Lopez, F. J. D., Bastein, T., & Tukker, A. (2019). Business model innovation for resource-efficiency, circularity and cleaner production: What 143 cases tell us. *Ecological Economics, 155*, 20–35.

Lorek, M. (2012). Eco-industrie et reconversion du territoire industriel et portuaire: le cas de Gdansk. *Marche et organisations* (2), 129–152.

Lundvall, B.-A. (1985). Product innovation and user-producer interaction. *The Learning Economy and the Economics of Hope, 19*, 19–60.

Martínez-Nicolás, J. J., Legua, P., Hernández, F., Martínez-Font, R., Giordani, E., & Melgarejo, P. (2021). Effect of phytoremediated port sediment as an agricultural medium for pomegranate cultivation: Mobility of contaminants in the plant. *Sustainability, 13*(17), 9661.

Mat, N. (2015). *Dynamiques de transition dans les territoires portuaires: apport de l'écologie industrielle et territoriale aux processus d'adaptation vers une société bas-carbone*. Saint-Etienne, EMSE.

Moore, J. (1996). *The death of competition: Leadership and strategy in the age of business ecosystems*. Book review and commentary by herb Rubenstein, President, Sustainable Business Group.

Moore, J. F. (2006). Business ecosystems and the view from the firm. *The Antitrust Bulletin, 51*(1), 31–75.

Nalebuff, B. J., Brandenburger, A., & Maulana, A. (1996). *Co-opetition*. HarperCollinsBusiness.

Nambisan, S., & Baron, R. A. (2013). Entrepreneurship in innovation ecosystems: Entrepreneurs' self–regulatory processes and their implications for new venture success. *Entrepreneurship Theory and Practice, 37*(5), 1071–1097.

OECD. (2018). *Oslo Manual*.

Pecqueur, B., & Zimmermann, J.-B. (2004). *Economie de proximités*. Hermes-Lavoisier.

Philbin, S. P., Jones, D., Brandon, N. P., & Hawkes, A. D. (2014). Exploring research institutes: Structures, functioning and typology. In *Proceedings of PICMET'14 conference: Portland International Center for Management of engineering and technology; Infrastructure and Service Integration*.

Pieroni, M. P., McAloone, T. C., & Pigosso, D. C. (2019). Business model innovation for circular economy and sustainability: A review of approaches. *Journal of Cleaner Production, 215*, 198–216.

Pintossi, N., Ikiz Kaya, D., & Pereira Roders, A. (2021). Assessing cultural heritage adaptive reuse practices: Multi-scale challenges and solutions in Rijeka. *Sustainability, 13*(7), 3603.

Promotion, D. (2021). *Des Fonds pour la Décarbonation de l'industrie et des Territoires*. https://www.dunkerquepromotion.org/2020/12/decarbonation-de-lindustrie-et-des-territoires/

Roberts, T., Williams, I., Preston, J., Clarke, N., Odum, M., & O'Gorman, S. (2021). A virtuous circle? Increasing local benefits from ports by adopting circular economy principles. *Sustainability, 13*(13), 7079.

Shahrokni, H., Årman, L., Lazarevic, D., Nilsson, A., & Brandt, N. (2015). Implementing smart urban metabolism in the Stockholm Royal Seaport: Smart city SRS. *Journal of Industrial Ecology, 19*(5), 917–929.

Shaw, D. R., & Allen, T. (2018). Studying innovation ecosystems using ecology theory. *Technological Forecasting and Social Change, 136*, 88–102.

Tansley, A. G. (1935). The use and abuse of vegetational concepts and terms. *Ecology, 16*(3), 284–307.

Torre, A. (2009). Retour sur la notion de proximité géographique. *Géographie, économie, société, 11*(1), 63–75.

Tsujimoto, M., Kajikawa, Y., Tomita, J., & Matsumoto, Y. (2018). A review of the ecosystem concept—Towards coherent ecosystem design. *Technological Forecasting and Social Change, 136*, 49–58.

Uche-Soria, M., & Rodríguez-Monroy, C. (2019). Solutions to marine pollution in Canary Islands' ports: Alternatives and optimization of energy management. *Resources, 8*(2), 59.

Urban planning agency Flandre-Dunkirk. (2020). *Les chiffres de la population décryptés - Région Flandre-Dunkerque*.

Uzunidis, D. (2010). Innovation et proximité. *La Revue des sciences de gestion* (1), 13–22.

Uzunidis, D. (2014). Entrepreneurship policies. *Entrepreneurship in Open Innovation Systems*.

Veyssière, S., Laperche, B., & Blanquart, C. (2021). Territorial development process based on the circular economy: A systematic literature review. *European Planning Studies*. https://doi.org/10.1080/09654313.2021.1873917

Volkmer Martins, B., Faccin, K., Espindula, E., & Balestrin, A. (2019). Understanding innovation ecosystems: A biomimetic approach.

Walrave, B., Talmar, M., Podoynitsyna, K. S., Romme, A. G. L., & Verbong, G. P. (2018). A multi-level perspective on innovation ecosystems for path-breaking innovation. *Technological Forecasting and Social Change, 136*, 103–113.

Williams, J. (2019). The circular regeneration of a seaport. *Sustainability, 11*(12), 3424.

WWF, W. W. F. f. N. (2014). *Living planet report 2014: Species and spaces, people and places gland.*

Yaghmaie, P., & Vanhaverbeke, W. (2019). Identifying and describing constituents of innovation ecosystems: A systematic review of the literature. *EuroMed Journal of Business*.

Circular Economy and Business Model Innovation: The Key Elements for a Sustainable Transition in Spain Circular Economy 2030

Yanina Kowszyk, Adeline Abou-Ali, and Lizbeth Arroyo

Abstract In recent years, the circular economy has arisen as a crucial driver for sustainable development across all fields and notably in the academia, industry, and policymaking. The adoption of alternative models that aim at reducing the use of resources, waste, and emissions in the European Union has therefore considerably increased. Accordingly, the development of the concepts at the scientific level has also advanced rapidly; however, the understanding of the mechanism of new theoretical frameworks that can shed light in regard to companies' practices is still missing. Thus, in this chapter, through a comparative analysis of six key sectors and companies' cases, we aim at applying a widely used theoretical framework to analyze circular business models, to recognize their characteristics while referring to Spanish SMEs, and to understand how these models are aligned with the main institutional instrument in the country: the Spanish 2030 Circular Economy Strategy.

1 Introduction

A growing societal pressure for more sustainable world has recently pushed countries and their governments to consider for the adoption of alternative models based on an ecological approach. In this sense, the circular economy (CE) has recently emerged as an essential perspective to a sustainable transition and has gained great importance on the agendas of policymakers (Brennan et al., 2015). This perspective has been reflected in the development of the Chinese Circular Economy Promotion Law in 2009 (Lieder & Rashid, 2016) and a comprehensive European circular economy package by the European Commission in 2015 (European Commission,

Y. Kowszyk (✉)
University of Groningen, Groningen, Netherlands

Universitat de Barcelona, Barcelona, Spain
e-mail: ykowszko7@alumnes.ub.edu

A. Abou-Ali · L. Arroyo
Universitat de Barcelona, Barcelona, Spain
e-mail: aabouaab7@alumnes.ub.edu; larroyo@ub.edu

V. Prokop et al. (eds.), *Business Models for the Circular Economy*, Sustainability and Innovation, https://doi.org/10.1007/978-3-031-08313-6_7

2015). Academia has also seen the rise of studies centered on the circular economy and its development as an important field of research with an increase in the number of articles and journals dealing with it during the last decade (Geissdoerfer et al., 2017). Moreover, companies are also becoming aware of the opportunities residing in the circular economy and have started looking for possible ways to make their business more eco-responsible by implementing new business models. Indeed, circular business models appear as an important parameter and are needed to optimize by closing the resource loops, as well as to deal with key resources and climate challenges (Bocken et al., 2018). While a growing literature on circular business models is emerging, the main theoretical literature frameworks for the circular business model innovation (CBMI) analysis have not yet been applied to Spanish small and medium enterprises (SMEs).

Thus, this chapter proposes the framework developed by Bocken et al. (2015, 2018) to help companies identify the key elements needed for a transition to a more sustainable business model, notably in the context of the circular economy strategic plan for 2030 adopted by the Spanish government. More specifically, this chapter seeks to answer the following research questions:

- What are the characteristics of the CBMI at the firm level in key selected SME sectors in Spain?
- How these circular business models are aligned with the main institutional framework in the country with respect to Spanish 2030 Circular Economy Strategy?

Our contribution follows the research agenda proposed by Centobelli et al. (2020) in addressing the influence of political initiatives and regulatory frameworks in the adoption of circular economy in a given country, in this case, Spain. The result of our study shed a light on the practical implications while implementing and transitioning to a more sustainable business model of six case studies in Spain and how their practices aligned with the regulatory framework proposed by Spain.

This chapter is structured as follows: First, we review the theoretical background on which the current research is drafted, followed by the context and methodology used, and finally, the paper concludes with terminating remarks on the contributions.

2 Concepts and Theory

The increasing relevance of sustainable development within governmental agendas is changing the rules of the game (Hall et al., 2010; Woolthuis et al., 2013). Policymakers' pursuit of sustainability transitions transforms existing structures, institutions, cultures, and practices toward more sustainable ones (Loorbach, 2010; Timmermans et al., 2014). Within the relevant actions included in the transition agenda, entrepreneurship plays a central role to solve sustainability issues through business activities and form a sustainable community (Hall et al., 2010). In fact, a key aspect to promote the development of CBMI is the establishment of normative regulations at the industry level, which depends on the dynamics and intensity of the interplay between policymakers and companies (Centobelli et al., 2020). This means that at a larger extent, drivers or barriers for the circular economy transition will depend on national/regional contextual factors, which are closely connected with the public policies (Centobelli et al., 2020).

In this context, the European Union (EU) has declared the need for academic research to face the environmental challenge and specifically consider topics related to circular economy (Urbinati et al., 2017). As a way to address the constant decline of the environment state, the circular economy approach has risen to the highest prominence in the last years (Lacy & Rutqvist, 2015). The reason behind this is the underlying fact that circular economy, as an alternative to linear economy, brings back environmental, social, and financial benefits (Lewandowski, 2016). With the aim of preventing raw materials to be converted into waste, circular economy enables companies to create and capture value from including the "R" principles into the business model: reusing, repairing, remanufacturing, or recycling (Centobelli et al., 2020). At the firm level, the new model is a set of strategic decisions aiming to preserve the embedded environmental and economic value of a product or service into a system (Centobelli et al., 2020).

In this chapter, the notion of CBMI is understood as innovating the business model to integrate, implement, and capitalize on circular economy practices (Bocken et al., 2018). Such practices might center on diverse aspects of the circular economy, as, for instance, product durability and design for product life extension, narrowing resource loops, and recycling initiatives to close the loop (Bocken et al., 2016).

However, as the topic is relatively new, the development of CBMI still faces several barriers at the firm level (Franco, 2017; Hopkinson et al., 2018) and claims for fundamental adaptations of industrial processes and business models (Ferasso et al., 2020; Vermunt et al., 2019; Manninen et al., 2018; Bocken et al., 2016). The influence of circular economy on business models has attracted maximum attention of major research scholars in recent years (Centobelli et al., 2020; Ferasso et al., 2020). With the support of case studies (Hopkinson et al., 2018), archetypes (Sassanelli et al., 2019), typologies (Vermunt et al., 2019; Bocken et al., 2016), and model comparison (Lewandowski, 2016), the literature have clarified certain facts. It clearly determines the circular business models, and strategies have, at least, the following determinants (Centobelli et al., 2020; Geissdoerfer et al., 2020; Lüdeke-Freund et al., 2019; Hopkinson et al., 2018; Clauss, 2017; Planning, 2015; Teece, 2010; Osterwalder & Pigneur, 2010):

- Firms need to secure proactive management providing strategic direction for long term. Managers are requested to believe, push the transformation of companies, and lead the engagement of all the value chain.
- The enlargement and adaptation of stakeholders' network, adjusting the organizational structure, and rethinking the relationship with the value chain partners are necessary. Although the organization-centric view is still predominant, the value network can bring the ecosystem view that can inform not only the new business model but also the new strategy.

- Backcasting, eco-design, and the role of digital technologies are key to enhance sustainable competitiveness and smart growth.
- Improving the management system and the operational level to deal with trade-offs, tensions, and opportunities will support the new business model.
- Marketing is needed to change the costumer culture – not only to value the innovation efforts of the market but also to accept remanufactured products instead of keep on buying new ones. The secondhand market claims for a new branding of reuse goods with a progressively growing educated consumer base.

While a circular economy business model can deliver significant improvements, it requires managers and practitioners to acquire new knowledge and to develop additional skills (Hopkinson et al., 2018) in order to engage into an experimental, iterative, and transformative process with learning cycles and sustainability checks (Bocken et al., 2018). This would be one of the main underlying challenges in managing circular business models (Hopkinson et al., 2018).

The iterative process includes different phases such as ideation, implementation, and evaluation and can result in various degrees of innovation, for example, a new activity added to a business model or a comprehensive change in various business model elements (Frankenberger et al., 2013). By rethinking how a company creates, delivers, and captures value, the framework for CBMI developed by Bocken et al. (2015, 2018) can be viewed as a holistic approach that has been referred to in the literature as a proven model for value creation with circular principles. In this chapter, the different components of the framework analysis for CBMI (Bocken et al., 2015, 2018) will be used to analyze a sample of SMEs in Spain and therefore are relevant to be considered:

(a) The *value proposition* "provides measurable ecological and social value in concert with economic value" (Bocken et al., 2014). Defining the right value proposition could imply experimentation on the model to effectively start the transition of the business toward a more sustainable model (Bocken et al., 2021). This requires the development of sustainability innovation and iteration capability, the involvement of internal stakeholders, and, finally, the external validation (Bocken et al., 2018). However, due to the lack of methods and tools for experimenting and testing models, the ongoing learning by doing process (Pieroni et al., 2019) and the role of ecosystem partnerships have become relevant (Geissdoerfer et al., 2020).
(b) The *value creation and delivery* represent how the firm is competing by demonstrating its way of doing business, describing its architecture, and defining its competitive advantage sources (Bocken et al., 2014). Therefore, compared with traditional business models, circular business model design requires a product systemic perspective (Bocken et al., 2016). Value creation includes maintenance of product and process, combination of resources and materials, purchasing upcycled waste, recycling of resources, dematerialization of products, and on-demand production process (Centobelli et al., 2020). Moreover, to ensure circular logistic loops, companies need to collaborate with different stakeholders allowing the maximization of products and material value (Bocken et al., 2018).

Hence, circular business models' value creation and delivery integrate multiple stakeholders in the innovation process, such as collaborators which might appear as different from value chain's conventional partners (Bocken et al., 2018), ranging from technological parks and innovation centers to innovative start-ups.

(c) The last element known as *value capture* refers to the revenue model. This means capitalizing additional revenue sources and intangibles, reducing costs, and changing their structure and achieving value preservation (Centobelli et al., 2020). An example of circular business model is the firm moving from selling only products to selling services (i.e., servitization), implying that initial investments will often be earned back after a gradual period. In this sense, it is important to bear in mind that it is still hard to find cost/benefit comparisons between linear and circular revenue models (Hopkinson et al., 2018). Therefore, considerable innovation will be needed in financial information and models to permit companies to transition successfully to circular business models (Bocken et al., 2014). Coherently, these financial models will need to be considered in the policy framework or in the industry voluntary standards to drive costs down (Hopkinson et al., 2018).

In this chapter, this framework is proposed to understand how companies create, propose, deliver, and capture value in each of the action lines provided by the Spanish 2030 Circular Economy Strategy, thus demonstrating what are the changes in the business models that are necessary to integrate the circular economy approach in accordance with main policy instruments in the field.

3 Methodology

3.1 Research Setting

This section has been developed in the context of the Spanish government's Circular Economy 2030 Strategy. This strategic plan aims at promoting "a new production and consumption model in which the value of products, materials and resources are maintained within the economy for as long as possible, with minimal waste and reusing as much as possible the waste that cannot be avoided. This strategy contributes to Spain's efforts to achieve a sustainable, decarbonized economy, which uses resources efficiently and is competitive. This strategy will be materialized in successive triennial action plans" (Circular Economy 2030, 2020).

For the design and implementation of the new CE 2030 Strategy, the High Commissioner has considered different frameworks at international, European, and national levels. Table 1 shows instruments and frameworks that have been considered to create the package of measures in regard to CE in the country.

From the presentation in December 2015 of the EU Action Plan for the Circular Economy in 2017, several legislative actions mainly linked to the field of waste have

Table 1 Frameworks adopted for CE transition in Spain

International	European	Spanish
The Paris Agreement (COP 21, 2015)	Europe 2020 strategy for smart, sustainable, and inclusive growth. Flagship initiative: "Roadmap to a resource-efficient Europe"	Spanish 2030 Circular Economy Strategy draft version (Ministry of Agriculture, Fishing, Nutrition and Environment and Ministry of Economy, Industry and Competitiveness)
2030 Agenda for Sustainable Development	7th Environment Action Programme (7th EAP), to make the EU a low-carbon, resource-efficient, environmentally friendly, and competitive economy	Law proposal on climate change and energy transition (Ministry for Energy Transition)
Sustainable Development Goals (SDG) 12	Circular Economy Action Plan. Package of measures for Circular Economy 2015 and 2018 including 54 actions on plastics, waste, chemical products, raw materials, and a monitoring plan	2030 Spanish Bioeconomy Strategy Horizon 2030 (Government of Spain)
New Urban Agenda—UN-Habitat	Circular Economy Action Plan. Circular Economy Action Package 2015 and 2018. With 54 actions covering the entire cycle. Among which there are measures on plastics, wastes, chemicals, critical raw materials, and a monitoring framework	The Spanish Action Plan to implement the Agenda 2030 toward and Spanish Strategy of Sustainable Development includes a special policy to increase and accelerate the Sustainable Development Goals (SDGs) adoption. The Circular Economy Local Strategy Model (FEMP) refers to circularity with crosscutting policies on new technology development and adoption, sustainable public purchasing, transparency, and participative governance. Spanish Urban Agenda, Ministry of Promotion, Macro Public Plan of Waste Management (PEMAR) 2016–2022

Adapted from COTEC Foundation (2019)

been developed. Table 2 illustrates several policy instruments for the CE promotion in Spain.

Based on the European Commission indicators used to assess progress toward a circular economy at the EU level, Spain adopts the same indicators but includes greenhouse gas emissions. These indicators aim at measuring the public policy application, the sustainability and circularity system adoption by the productive sector, and the consumer's choice of products and services considering sustainability criteria (Circular Economy 2030, 2020). Hence, the government framework for follow-up, monitoring, and assessment is based on ten indicators grouped in four stages and aspects of circular economy: (1) producers and consumers, (2) waste management, (3) secondary raw materials, and (4) competitiveness and innovation (Circular Economy 2030, 2020).

Table 2 Policy actions toward CE in Spain (2017–2020)

Policy actions	Government level	Institutions responsible/leadership
2030 Circular Economy Strategy	National	Ministry of Agriculture, Fishing, Nutrition and Environment and Ministry of Economy, Industry and Competitiveness, in collaboration with other ministries, the autonomous communities, and the Spanish Federation of Municipalities and Provinces (FEMP)
"Circular Economy Pact" (2017)	National	Ministry of Agriculture, Fishing, Nutrition and Environment
Law proposal on climate change and energy transition (draft submitted in 2018)	National	Parliament
Contracts law	National	Parliament
"Green and Circular Economy Strategy"	Regional	Catalunya
Circular Economy Strategy Euskadi 2030	Regional	Euskadi
Andalusian Sustainable Development Strategy 2030	Regional	Andalusia

Adapted from COTEC Foundation (2019)

Furthermore, the Circular Economy 2030 Plan includes the monitoring of six key sectors including construction; farming, fishing, and forestry; industry in general; consumption goods; tourism; and textiles and garment in the following action lines (Circular Economy 2030, 2020):

- *Production*: "Conception, design, and manufacturing of a product that is easier to repair, with a longer useful life, which may be updated and which, at the end of its useful life, creates less waste or even recyclable (and of course free of harmful substances)."
- *Consumption*: "Reverting the current trend towards excessive consumption and gearing towards a more conscious consumption model that includes accessible services; this is a precondition to further prevention and reduction of waste, and, when appropriate, to promote adequate recycling."
- *Waste management*: "In a worldwide context in which raw materials are increasingly scarce and expensive, recycling just 37.1% of waste (the current rate in Spain) is, in itself, a waste of available resources; urgent action must be taken with regard to recovering and recycling."
- *Secondary raw materials*: "Using secondary raw materials allows making more sustainable use of natural resources, as well as enabling consumers to trust more conscious consumption models."
- *Water reuse and purification*: "This factor is included as an individual factor due to the importance of water in the Iberian Peninsula. Acknowledging its fundamental nature, it was decided to consider this aspect separately, beyond obtaining

secondary raw materials, in recognition of its special significance in the Spanish economy and our leading position in water reuse."

- *Awareness raising and participation*: "Due to the special relevance of citizen's involvement in the progression towards a circular economy. On one hand, since citizens make decisions with regard to the products and services, they need to have information about them: an informed and conscious consumption is the only way towards waste hierarchy with a focus on prevention. On the other hand, domestic waste separation at the origin, both in households and in the service sector, is everyone's responsibility. Therefore, it is fundamental to involve the entire society so waste separation is carried out adequately facilitating better recycling practices which allows reaching the European goals."
- *Research, innovation and competitiveness*: "Research, innovation and competitiveness policies are core to the Strategy, so it made sense to address them in a separate section."
- *Employment and training*: "Training, development of new capacities adapted to new opportunities, training oriented to jobs arising from the transition towards a circular economy, job creation and improvement of already existing jobs require special policies, which are to play a very important role in the future Circular Economy."

Beyond the public efforts and measures, previous reports (COTEC, 2017) concluded that circular economy landscape in Spain and the initiatives in this area are still incipient and that the measures adopted are mainly focused on downstream environmental policies such as management of waste, increased recycling, and reduction of landfills, supported by initiatives such as the Macro Public Plan of Waste Management (PEMAR, 2016–2022). In regard to the firm's adoption of the CE model, there is still a lack of uniformity in the measurement of practices (Barón et al., 2020), and companies are still focused on complying with the law, cost savings, material provision, and resource utilization and building a new image (Ormazabal et al., 2018), instead of embedding a holistic perspective.

3.2 Research Approach and Case Selection

In this chapter, we adopted a qualitative approach, by realizing a comparative case study analysis. The strength of this research approach relies on the analysis of several cases to gain an in-depth understanding and insights from the phenomenon (Yin et al., 2018). An explorative cross-case analysis allows us to study the similarities and differences between the six cases selected and that share a common institutional business framework but developed different CBMI adoption strategies (Alshawaaf & Lee, 2021). We applied a content analysis for data collection purposes, using secondary data source (Pieroni et al., 2020). Evidencing the interaction between institutions and companies, in relation with the principles of the Circular Economy 2030 Plan, we identified the key elements used by companies to transition toward

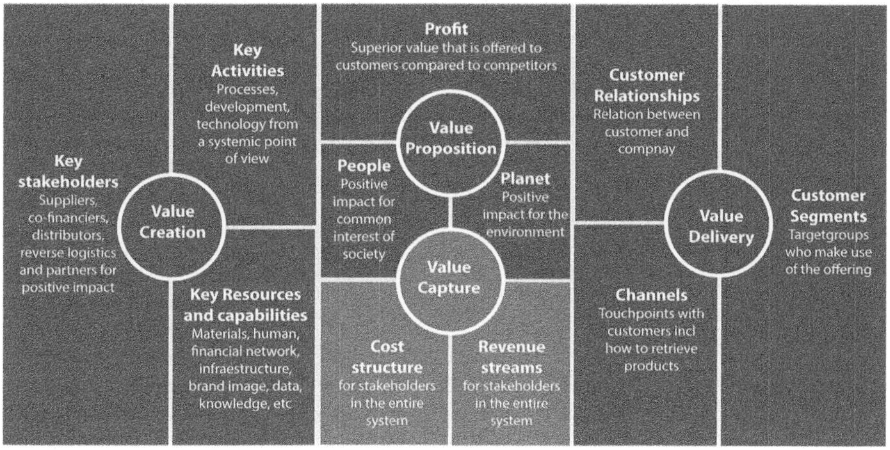

Fig. 1 Sustainable BM canvas. Adopted from Bocken et al. (2018)

circular business models (Bocken et al., 2018). The sampling process was based on specific criteria previously applied in the literature reviewed to develop this chapter. We here after describe more precisely the selection criteria applied to companies. First, the companies sampled must include different levels of maturity of transition to a sustainable business model, in addition of different companies' size and industry, in order to ensure the validity and reliability of the sample (Bocken et al., 2018). Second, the companies must be SMEs that are or have been innovating their BMs to a more sustainable one, with the SME size including start-ups, microenterprises, and small- or medium-sized enterprises, given the lack of representation of such companies in the current CE and sustainable BM literature (Heikkilä et al., 2018). Third, companies must develop their transition to a CBM in the Spanish context of the Circular Economy 2030 Plan, as this research aims at showing a light on firm transition to a circular business model (CBM) in this specific institutional setting. Fourth, the BM that resulted from this transition had to be new to the firm, so that the CBMI was both highly present and clearly observable through the data collected (Pettigrew, 1990). Finally, the CBMI also had to be relatively new to its corresponding sector, representing a challenging opportunity, with uncertainty and few chances to be imitated by other firms (Rivkin, 2000).

An initial sample of 28 companies was defined, on which the selection criteria previously listed were applied, resulting in a final sample of six cases. Each case represents one of the economic sectors mentioned by the Circular Economy 2030 Plan.

The companies' cases were mapped in accordance with the sustainable business model value proposition, value creation delivery, and value capture framework recognized in the circular and sustainable business model literature (Osterwalder & Pigneur, 2010; Bocken et al., 2013, 2014, 2016, 2018) and more particularly following the sustainable business model (BM) canvas developed by Bocken et al. (2015) (Fig. 1).

Table 3 Description of selected six case study companies

Company name	Founding date	Location	Size	Industry	Activities
O11h	2020	Catalunya	Between 11 and 50	Construction	Creating a new type of construction company based on the latest advances in design, technology, industrialization, sustainability, and health to deliver buildings faster, cheaper, more reliably, and in a way that is both people-friendly and environmentally friendly
Debuencafé	2015	Several Spanish cities	Between 1 and 10	Farming, fishing, and foresting	Producing coffee in a sustainable way from production to consumption. Their coffee in capsules is 100% organic. The organic agriculture certification guarantees that their coffee takes care of the environment
Ekorec	2012	País Vasco	120	Industry	Recycling, I & D, turnkey contracts, clean energy generation, water treatment
Zicla	2005	Catalunya	Between 11 and 50	Consumer goods	Accessible, inclusive, and friendly streets with small environmental footprint
ECO-ONE	2020	Valencia	Between 1 and 10	Tourism	First company that proposes this sustainable service in the tourism industry. The company defends that sustainability goes hand in hand with profitability and believe that all hotels can be more sustainable by changing their habits
Ecoalf	2009	Several Spanish cities	Between 100 and 500	Textile and garment	New generations will buy recycled clothing because they are aware that in 2050, we will have more plastic than fish in the ocean

We review the companies' webpages, social media, published interviews of companies' owners, sustainability experts, and consultants highlighting the value of these case studies and documents available online referring to them. Data collection was done during September/October 2021 by the authors who have qualitative analysis knowledge and experience and had worked with digital ethnography in the past.

Table 3 shows the company cases selected and their characteristics. We, hereafter, provide a short description of each case selected.

O11h Founded in 2020 in Barcelona, O11h is a construction company revolutionizing the sector through technology to create intelligent, high-quality, healthy, and sustainable buildings. It offers a turnkey construction service that includes design (architecture and engineering), procurement, manufacturing, construction, and digital support for optimal property operation and maintenance.

Debuencafé Born in Madrid in 2015, Debuencafé is a Spanish coffee company, which produces its coffee in a sustainable way from production to consumption. Their coffee, sold in compostable capsules, is 100% organic and Nespresso-compatible. The company seeks to become the flagship company for sustainability in the world of coffee capsules.

Ekorec Located in Andoain, Basque Country, Ekorec is a reference in plastic's circular economy within Spain, especially in PET materials. Since 2012, it is specialized in recycling PET bottles using postconsumer and postindustrial waste. The resulting flakes (color and transparent) are transformed into PET granule, agglomerate, and PET sheet. The company's customers then use those materials to produce new packaging such us PET trays and bottles.

Zicla Since 2005, the Barcelona-based company has supported cities in their transformation process needed so they can meet the challenges of the twenty-first century. The company works toward making cycling mobility safer and cities more accessible to all citizens by creating products manufactured using the waste generated by the cities themselves.

ECO-ONE Founded in 2020, the Valencia-based company has for main objective to help hotels on their path to sustainability under a single interlocutor, always recommending profitability and sustainability measures to the hotelier. Its vision rests on the belief that sustainability goes hand in hand with profitability and that all hotels can be more sustainable by changing their habits. Hence, the company has become hoteliers' free external sustainability agency.

Ecoalf Created in 2009, Ecoalf is a Madrid-based company specialized in stopping the careless use of natural resources by creating the first generation of recycled products possessing the same quality and design as the best non-recycled products on the fashion market.

4 Results

The analysis of the cases started with framing the circularity of each business model, which led to the definition and exploration of the different CBMI components for each company studied shown in Table 4. The results are organized using the literature framework of value proposition, value creation, value delivery, and value capture, followed by how the components of these CBMI are aligned with the institutional Spanish Circular Economy 2030 Plan.

Table 4 Exploration of the circular business model innovation components of each case studied

Company name	Value proposition	Value creation	Value delivery	Value capture
O11h	Creating a new type of construction company based on the latest advances in design, technology, industrialization, sustainability, and health to deliver building faster, cheaper, more reliable, and in a way which is both people-friendly and environmentally friendly	First company to offer such sustainable turnkey construction process and building in Spain	Providing turnkey sustainable projects to customer	One-stop shop model
Debuencafé	Producing coffee in a sustainable way from production to consumption. Their coffee in capsules is 100% organic. The organic agriculture certification guarantees that their coffee takes care of the environment	B Corp producing coffee in a sustainable way, their product coming from the Earth, and once consumed and used going back to Earth	Online, onsite, and in some hotels, restaurants, bar, etc. which are partners	Payment by products
Ekorec	Recycling, I & D, turnkey contracts, clean energy generation, and water treatment	Recycling with energy clean generation and water treatment allows the company to reach its customers with a high-quality product, which can be prepared and processed by customizing to clients' needs	Focused relations with specific segments, creating products for their needs	Saving costs from energy and water consumption, payments made by customers
Zicla	Accessible, inclusive, and friendly streets with small environmental footprint	It caters to more accessible, inclusive, friendly, sustainable cities, where waste is brought back in the form of products that help solve their problems. Safe	Studying, visiting, economic alternative analysis, project presentation, technical assistance, and monitoring	Payments for services and products

(continued)

Table 4 (continued)

Company name	Value proposition	Value creation	Value delivery	Value capture
		urban cycling and improvement of urban accessibility in general and accessibility at bus stops in particular		
ECO-ONE	First company that proposes this sustainable service in the tourism industry. The company defends that sustainability goes hand in hand with profitability and believe that all hotels can be more sustainable by changing their habits	The company provides hotels with a turnkey solution for a sustainable transition including eco-management, eco-design, eco-efficiency, and eco-solution	Accompanying hotels in their "eco-transition" through sustainable services offers	Service model (payment by service)
Ecoalf	New generations will buy recycled clothing because they are aware that in 2050, we will have more plastic than fish in the ocean	From PET bottles to a 30-year jacket. From waste collection to a circular fishing industry. Products have the slogan "there is not planet b"	Clear target, strategy of customer relationship focused in sustainability. Clear and great number of channels	Payment by products

4.1 Circular Value Proposition

We found that all companies in our sample have introduced different *innovations in their value proposition*:

- *O11h* is disrupting the traditional and long-time stagnating construction industry by developing a value proposition based on providing a more sustainable, faster, cheaper, and more reliable way of constructing buildings using specific eco-friendly materials such as wood and ensuring a carbon neutral process.
- *Debuencafé* is changing the traditional farming, fishing, and foresting industry linear value proposition by offering its customer a sustainable coffee ensuring a sustainable process from cultivation to encapsulation and by reincorporating its products directly back into the nature.
- *Ecorec* is challenging the classical use of plastic in the traditional industry value proposition, by recycling PET bottles using postconsumer and postindustrial

waste allowing for its customers to use the new material outcome to produce new packaging such us PET trays and bottles.

- Looking at the consumer goods industry, *Zicla* value proposition introduces waste as a source to solve problems in cities accessibility, by creating products manufactured using the waste generated by the cities themselves.
- The tourism industry, similar to the construction one, is still based on a very traditional value proposition offering classical hotel room products and services. However, *ECO-ONE* is challenging this traditional value proposition by offering hotels to help them in their sustainable transition, proposing them eco-friendly alternatives.
- Finally, the traditional fashion and garment industry value proposition of offering linearly produced products is being reconsidered, notably by companies such as *Ecoalf*, who developed a value proposition based on providing their customers with clothing produces out of ocean plastics.

4.2 Circular Value Creation

The companies studied show the *development of specific key activities* in terms of process development and technologies, which constitute a source of competitive advantage:

(a) Developing research and innovation processes to find the solutions to issues such as material treatment, recycling, manufacturing practices, new product delivery (e.g., plastic flakes, low-impact fibers), traffic segregation objects and lines, and management and operational systems for users and clients
(b) Mapping nature-based risks and opportunities throughout the value chain to understand how to address them and make the business model transition financially feasible and sustainable
(c) Standardizing innovative and circular processes such as construction, farming, or recycling process, including predesign, parametrization, monitoring, facilitation, supporting implementers, and customer returning process, among others
(d) Using technological and clean components or materials such as certified wood and biodegradable, recyclable, nontoxic, and nonchemical materials
(e) Using renewable and clean energy sources and implementing operational practices to protect the atmosphere
(f) Integrating digitalization in the production, manufacturing, design, and retail system
(g) Manufacturing and assembling customized items for clients and users
(h) Measuring, testing, and controlling quality, by checking for potential errors implemented

We also observed *specific resources and skills that contribute to the value creation* in our sample:

(a) Digitalization and modeling from design to operation and maintenance, special engineering, and farming abilities for ecological production, monitoring, evaluation, and certification skills
(b) Research and development, evidence mapping, knowledge about trends in sustainable cities, health threat identification, new materials, processes, management systems, and emerging natural-related energy sources
(c) Financial management, branding, and marketing
(d) Partnership scheme understanding and networking know-how, advocating, and campaigning

4.3 Circular Value Delivery

We found a *diversity of strategies* developed by companies to deliver value, ranging from turnkey projects and accompanying other businesses in their sustainable transition to offering standardized and customized products online and onsite:

- At the customer segment level, the companies studied targeted specific customers that are aware and interested in sustainable projects or products. Some examples of customer targeted include city halls, real estate and hotel operators, and consumers of sustainable products at industrial scale (such as manufacturers) or at the consumer level (such as sustainable conscious buyers at the national or international level).
- In terms of customer relations and touchpoints, the companies appeared to use online (via website, platforms, social media, TED talks, videos, campaigns) or through onsite sales.

4.4 Circular Value Capture

At the cost structure level, the companies studied showed their way of investing in R & D to develop sustainable production processes or acquire specific technologies, as well as hiring skilled and specialized workforce to achieve their circular goals.

Regarding the companies' *revenue models*, we found that traditional revenue streams where still used, such as the "payment for a product" model, but also that new revenue schemes emerged in order for a transition into a CBM. These models are the "service model" which is implying payments for services reflecting a certain servitization while transitioning to CBM and the "one-stop shop" model which aims at providing the customers with all the products essential aspects so that it does not need to look for anything elsewhere and which brings along energy, water, and resource usage savings altogether.

4.5 Alignment with Spain Circular Economy 2030 Plan

In this section, Table 5 summarizes how each company includes actions in line with the Spanish government in its Circular Economy 2030 Plan action lines.

According to this comparative perspective of the cases:

- All companies shared the use of low-impact materials, the standardization of their processes, the reduction of emissions, and R & D activities.
- Most of the companies also had local business partners, implemented recycling practices, and customized their product/services or offer turnkey projects.
- Just one company from the textile and garment industry implemented upcycling processes, and two have international certifications, one from the farming, fishing, and foresting industry and the other one from the textile and garment sector.

In terms of economic sector:

- Our case in the construction sector showed the industrialization of components and the digitalization of the design, building, operation, and maintenance processes.
- In the farming, fishing, and foresting, we identified nature-positive practices and a 100% organic production process. In this case, it is interesting to note the fact that due to the nature of its activity, the company has innovated to include and support groups in social risk.
- In the industry sector, we can report practices of clean energy and water treatment installed.
- In the consumer goods sector, the company in our sample is committed to solve long-term citizen problems along with the city's government by providing solutions in major current city hurdles such as accessibility, traffic, and mobility.
- In the case of tourism, the company studied is committed to create a behavioral change toward a sustainability transition in hotels.
- In the textile and garment sector, the company studied uses several touchpoints with its clients to attract the ones who are part of a generation aware of climate change. In this case, the company's behavior is similar to its customers, meaning that the company is a key activist in nature protection.

5 Discussion

In this chapter, we applied a framework for CBMI analysis to the study of six Spanish companies. In our sample, we did not find circular business model innovation addressing the entire meaning of the term; however, insights about particular strategies and actions reflect an incipient development of this type of BM in Spain. Moreover, some insights can be gained from approaches such as the lean start-up (Ries, 2011) and recent research on circular business model experimentation (Bocken et al., 2014, 2018).

Table 5 Case alignment with Spain Circular Economy 2030 Plan action lines

Company	Sector	Circular specialization	CE 2030 Plan action lines							
			Production	Consumption	Waste management	Secondary raw materials	Waste reuse and purification	Awareness raising and participation	Research, innovation, and competitiveness	Training and development
O11H	Construction	Digital design, building process, operation, and maintenance. Industrialization of components	√	√	√	√	√	√	√	√
Debuencafé	Farming, fishing, and foresting	100% organic and nature positive. Supporting social risk groups	√	√	√	√	√	√	√	√
Ekorec	Industry	Clean energy generation. Water treatment	√	√	√	√		√	√	√
Zicla	Consumer goods	City problem solution as a service and product: accessibility, traffic, and mobility. Governmental support required	√	√	√	√	√	√	√	
ECO-ONE	Tourism	Habits, routines, and business practice changer. External sustainability agency. Distinctive revenue model with suppliers instead of clients	√	√	√	√		√	√	
Ecoalf	Textile and garment	Environmental advocator and activist. Focused on young audience	√	√	√	√	√	√	√	√

As observed in the companies' value proposition section, most of the companies are looking for having a positive impact on the environment and society as a whole, joining previous research findings (Bocken et al., 2014, 2018). Additionally, our sample confirms the published results of Pieroni et al. (2019) referring to the fact that companies are learning by implementing and testing new business models. This includes raising awareness among their customers and suppliers by advertising their action and value to differentiate from their competitors though highlighting environmental and social positive impact.

With respect to environment, we found companies in our sample using eco-friendly materials (e.g., wood), recycling supplies (e.g., PET bottles and other plastic wastes), developing sustainable processes (e.g., zero emission, coffee cultivation, or hotel management system), and reducing CO_2 emissions (e.g., saving water and energy, reducing the use of fossil fuel energy consumption). We also observed a decline in the amount of generated or existing waste, dropping the plastic and artificial fiber consumption by offering recycled plastic and introducing new products or services being more sustainable.

To ensure a positive impact on the society, the companies in our sample implemented actions, which are listed as follows:

1. Eliminating toxic materials in construction to ensure healthy infrastructure (e.g., O11h)
2. Supporting groups at risk of social exclusion and buying from national suppliers (e.g., making citizen mobility and city accessibility more inclusive (e.g., Zicla))
3. Reducing the ecological footprint of tourism (e.g., ECO-ONE) and promoting gender equality, an altruism culture, and more responsible value chains

With regard to value creation, the main capabilities identified in our sample are digitalization and modeling, research and development, branding, marketing, and the ability to create partnership schemes (Geissdoerfer et al., 2020). The firms are especially conscious about the usage of water, as a key resource, during their production process, which is aligned to the water reuse and purification action aim in the Circular Economy 2030 Plan.

Moreover, confirming the previous findings (Osterwalder & Pigneur, 2010; Bocken et al., 2018; Velter et al., 2020), specific stakeholders were recognized as key collaborators to help companies create the value for their sustainable business model. This includes industrial partners, international organizations, research and innovation centers, testing labs, think tanks, business associations, certification entities, health organizations, local farmers, professionals, waste collectors, fishermen, suppliers and distributors, impact investors, banks, city mayors, and councils, among others.

At the value delivery level, companies should develop customer relationship that goes beyond the linear model of selling product and services by building a collaborating relationship and by targeting specific customer segment which are aware and interested in sustainable projects or products. Companies should also use specific touchpoints such as online advertising to reach their sustainable goal and customer target, as observed in sustainable business model research (Bocken et al., 2018; Velter et al., 2020).

A best practice in terms of creating specific customer touchpoint design is Zicla. In this case, far from a retrieving model, the goal is building a long-term collaboration-based relation with their clients, in this case, city halls, to solve emerging problems that will create new business opportunities in a collaborative creation process.

Another best practice is the case of Ecoalf, which describes in detail the product retrieving process for their customers, notably including its environmental implications, by disclosing not only the economic results of the process but also the environmental one.

A missing piece of information in our sample refers to transportation, which could be indicating a key challenge while designing a CBMI. In this sense, shipping and land transport seemed to be a hurdle in the process of creating a more sustainable approach.

Finally, regarding the value capture dimension, companies in our sample are investing in R & D focusing exclusively on process innovation by using new technologies and by employing and training their collaborators, which is also reflected in previous literature findings (Bocken et al., 2018; Geissdoerfer et al., 2020), as well as specified in action lines of Spanish government.

As previously discussed in the result section, the CBMI characteristics of our sample are aligned with the Spanish Circular 2030 Strategy. In the company set, there are some best practices to acknowledge:

- At the production level, remarkable eco-friendly product developments are represented in the cases of *Debuencafé*, *O11h*, and *Ecoalf*.
- At the consumption level, most of the companies in this chapter control their own material consumption by ensuring reverse logistics making their customers use the product or service without generating waste.
- The waste management action line is well put in place by companies such as *Zicla*, *Ecoalf*, or *Debuencafé* whose products come from recycled product that can then be again recycled in turn by their customers.
- Companies such as *Zicla*, *Ecoalf*, *Debuencafé*, *ECO-ONE*, and *O11h* understand correctly how to use secondary raw materials and ensure a sustainable use of natural resources.

6 Conclusion

In this chapter, we applied the CBMI developed by Bocken et al. (2014, 2018) to six case studies in Spain to understand the main characteristics of their business models and identify their alignment with the principal institutional framework in the country: the 2030 Circular Economy Strategy.

From our findings, we identified a key challenge in consumer education, marketing, and branding. Value capture for business depends on these factors, and actions in this sense are isolated and driven individually by the companies in the sample.

On the other hand, there may be a progressive need for companies and policymakers to adapt and follow technological advances due to the potential high-tech competition led by big data and the use of artificial intelligence and its impact in cost reduction and potential redefinition of market players and leaders. In this sense, companies do not appear to be reacting together but separately, and although policies refer to digitalization, this seems far from building a solid country position in terms of technology acquisition and investment in key economic sectors. Therefore, more private-public collaborations to face technological challenges as well as consumer awareness seem to be a priority.

The insights brought by this chapter can be used to compare the impact of circular economy strategies implemented in each European country. In the case of Spain, recycling and waste management are significant in the policy context; however, this is not the case when refereeing to water use, transports, and logistics. National and regional instruments are necessary to support industrial efforts if the development of circular value chains is a policy goal.

Future research can explore and focus on the coordination between business- and policy-driven CBM development mechanisms and elements. In addition, further investigation can focus on how institutions could measure the efficiency of their sustainable policy through CBMI. Finally, future research could also focus on how CBMI processes influence positive social impact.

References

Alshawaaf, N., & Lee, S. H. (2021). Business model innovation through digitisation in social purpose organisations: A comparative analysis of Tate Modern and Pompidou Centre. *Journal of Business Research, 125*, 597–608.

Barón, A., de Castro, R., & Giménez, G. (2020). Circular economy practices among industrial EMAS-registered SMEs in Spain. *Sustainability, 12*(21), 9011.

Bocken, N., Short, S., Rana, P., & Evans, S. (2013). A value mapping tool for sustainable business modelling. *Corporate Governance*.

Bocken, N. M. P., Rana, P., & Short, S. W. (2015). Value mapping for sustainable business thinking. *Journal of Industrial and Production Engineering, 32*(1), 67–81.

Bocken, N., Short, S., Rana, P., & Evans, S. (2014). A literature and practice review to develop sustainable business model archetypes. *Journal of Cleaner Production, 65*, 42–56.

Bocken, N. M. P., de Pauw, I., Bakker, C., & van der Grinten, B. (2016). Product design and business model strategies for a circular economy. *Journal of Industrial and Production Engineering, 33*, 308–320.

Bocken, N. M. P., Schuit, C. S. C., & Kraaijenhagen, C. (2018). Experimenting with a circular business model: Lessons from eight cases. *Environmental Innovation and Societal Transitions, 28*, 79–95.

Bocken, N. M., Weissbrod, I., & Antikainen, M. (2021). Business model experimentation for the circular economy: Definition and approaches. *Circular Economy and Sustainability, 1*(1), 49–81.

Brennan, G., Tennant, M., & Blomsma, F. (2015). Business and production solutions: Closing the loop. In H. Kopnina & E. Shoreman-Ouimet (Eds.), *Sustainability: Key issues* (pp. 219–239). Earth Scan, Routledge.

Centobelli, P., Cerchioni, R., Chiaroni, D., Del Vecchio, P., & Urbaniti, A. (2020). Designing business models in circular economy: A systematic literature review and research agenda. *Business Strategy & the Environment, 29*, 1734–1749.

Circular Economy 2030. (2020). Retrieved March 14, 2022, from chrome-extension://efaidnbmnnnibpcajpcglclefindmkaj/viewer.html?pdfurl=https%3A%2F%2Fwww.miteco.gob.es%2Ffr%2Fcalidad-y-evaluacion-ambiental%2Ftemas%2Feconomia-circular%2F200714eeec_resumenejecutivo_en_tcm36–510578.pdf & clen=1358149 & chunk=true

Clauss, Y. (2017). Measuring business model innovation: Conceptualization, scale development, and proof of performance. *R & D Management, 47*(3), 385–403.

Cotec Foundation. (2017). *La situación y evaluación de la economía circular en España*. Retrieved March 27, 2022, from https://cotec.es/observacion/la-economia-circular-2017/a34f8b7e-88ab-4640-b242-68f6b352169a

Cotec Foundation. (2019). *Situation of the Circular Economy in Spain*. Retrieved March 27, 2022, from https://cotec.es/observacion/economia-circular-2019/37040c86-20dc-4a51-8b3b-568994888cb6#:~:text=La%20econom%C3%ADa%20circular%20ofrece%20una,de%20fabricaci%C3%B3n%2C%20distribuci%C3%B3n%20y%20consumo

European Commission. (2015). Closing the loop – An EU action plan for the Circular Economy, Com (2015) 614 communication from the commission to the European parliament, the council, the European economic and social committee and the committee of the regions. European Commission, Brussels. Retrieved March 14, 2022, from https://www.eea.europa.eu/policy-documents/com-2015-0614-final

Ferasso, M., Beliaeva, T., Kraus, S., Clauss, T., & Ribeiro-Soriano, D. (2020). Circular economy business models: The state of research and avenues ahead. *Business Strategy and the Environment, 29*(8), 3006–3024.

Franco, M. A. (2017). Circular economy at the micro level: A dynamic view of incumbents' struggles and challenges in the textile industry. *Journal of Cleaner Production, 168*, 833–845.

Frankenberger, K., Weiblen, T., Csik, M., & Gassmann, O. (2013). The 4I-framework of business model innovation: A structured view on process phases and challenges. *International Journal of Product Development, 18*, 249–273.

Geissdoerfer, M., Savaget, P., Bocken, N. M. P., & Hultink, E. J. (2017). The circular economy – A new sustainability paradigm? *Journal of Cleaner Production, 143*, 757–768.

Geissdoerfer, M., Pieroni, M. P., Pigosso, D. C., & Soufani, K. (2020). Circular business models: A review. *Journal of Cleaner Production, 123741*.

Hall, J. K., Daneke, G. A., & Lenox, M. J. (2010). Sustainable development and entrepreneurship: Past contributions and future directions. *Journal of Business Venturing, 25*(5), 439–448.

Heikkilä, M., Bouwman, H., & Heikkilä, J. (2018). From strategic goals to business model innovation paths: An exploratory study. *Journal of Small Business and Enterprise Development, 25*(1), 107–128. https://doi.org/10.1108/JSBED-03-2017-0097

Hopkinson, P., Zils, M., Hawkins, P., & Roper, S. (2018). Managing a complex global circular economy business model: Opportunities and challenges. *California Management Review, 60*(3), 71–94.

Lacy, P., & Rutqvist, J. (2015). *Waste to wealth: The circular economy advantage*. Springer.

Lewandowski, M. (2016). Designing the business models for circular economy—Towards the conceptual framework. *Sustainability, 8*(1), 1–28, 43.

Lieder, M., & Rashid, A. (2016). Towards circular economy implementation: A comprehensive review in context of manufacturing industry. *Journal of Cleaner Production, 115*, 36–51.

Loorbach, D. (2010). Transition management for sustainable development: A prescriptive, complexity-based governance framework. *Governance, 23*(1), 161–183.

Lüdeke-Freund, F., Gold, S., & Bocken, N. M. (2019). A review and typology of circular economy business model patterns. *Journal of Industrial Ecology, 23*(1), 36–61.

Manninen, K., Koskela, S., Antikainen, R., Bocken, N., Dahlbo, H., & Aminoff, A. (2018). Do circular economy business models capture intended environmental value propositions? *Journal of Cleaner Production, 171*, 413–422.

Ormazabal, M., Prieto-Sandoval, V., Puga-Leal, R., & Jaca, C. (2018). Circular economy in Spanish SMEs: Challenges and opportunities. *Journal of Cleaner Production, 185*, 157–167.

Osterwalder, A., & Pigneur, Y. (2010). *Business model generation: A handbook for visionaries, game changers, and challengers*. Wiley.

Pettigrew, A. M. (1990). Longitudinal field research on change: Theory and practice. *Organization Science, 1*(3), 267–292.

Pieroni, M. P., McAloone, T. C., & Pigosso, D. C. (2019). Business model innovation for circular economy and sustainability: A review of approaches. *Journal of Cleaner Production, 215*, 198–216.

Pieroni, M. P., McAloone, T. C., & Pigosso, D. C. (2020). From theory to practice: Systematising and testing business model archetypes for circular economy. *Resources, Conservation and Recycling, 162*, 105029.

Rivkin, J. W. (2000). Imitation of complex strategies. *Management Science, 46*(6), 824–844.

Teece, D. J. (2010). Business models, business strategy and innovation. *Long Range Planning, 43*(2–3), 172–194.

Timmermans, J., Van Der Heiden, S., & Born, M. P. (2014). Policy entrepreneurs in sustainability transitions: Their personality and leadership profiles assessed. *Environmental Innovation and Societal Transitions, 13*, 96–108.

Urbinati, A., Chiaroni, D., & Chiesa, V. (2017). Towards a new taxonomy of circular economy business models. *Journal of Cleaner Production, 168*, 487–498.

Velter, M. G. E., Bitzer, V., Bocken, N. M. P., & Kemp, R. (2020). Sustainable business model innovation: The role of boundary work for multi-stakeholder alignment. *Journal of Cleaner Production, 247*, 119497.

Vermunt, D. A., Negro, S. O., Verweij, P. A., Kuppens, D. V., & Hekkert, M. P. (2019). Exploring barriers to implementing different circular business models. *Journal of Cleaner Production, 222*, 891–902.

Woolthuis, R. K., Hooimeijer, F., Bossink, B., Mulder, G., & Brouwer, J. (2013). Institutional entrepreneurship in sustainable urban development: Dutch successes as inspiration for transformation. *Journal of Cleaner Production, 50*, 91–100.

Yin, R. K., Campbell, D. T., & Donald, T. (2018). *Case study research and applications: Design and methods*. Sage.

Part III
Central and Northeast European Perspective

Green Tech for Green Growth? Insights from Nordic Environmental Innovation

Massimo Pizzol and Mikael Skou Andersen

Abstract Under a green growth strategy, technological solutions play a key role in decoupling growth and emissions—but bringing new technologies to the market remains difficult. We investigate here based on insights from the Nordic countries, often considered to be leading in environmental innovation, how and why the journey of many green technologies from their early conception to their large-scale adoption and upscaling remains a serious challenge. Our research is based on a sample of green patents complemented by interviews with their inventors and patentees. We observe that green technology innovation generally comprises seven different stages: R & D, start-up, patenting, upscaling, dealing with risk, commercialization, and adoption. Four main factors influence and shape these stages: the organizational setting, the technology-specific characteristics, the funding available, and the commercial market conditions. The insights from our study on green technology development allow for distinguishing shared framework conditions from technology-specific ones and can help inform innovation policies. Upscaling of green technology innovations remains an almost impervious and difficult process, harder and more enduring than most inventors—and certainly the public as well as many policymakers—expect.

Supplementary Information The online version contains supplementary material available at [https://doi.org/10.1007/978-3-031-08313-6_8].

M. Pizzol
Department of Planning, Aalborg University, Aalborg, Denmark
e-mail: massimo@plan.aau.dk

M. S. Andersen (✉)
Department of Environmental Science, Aarhus University, Aarhus C, Denmark
e-mail: msa@envs.au.dk

V. Prokop et al. (eds.), *Business Models for the Circular Economy*, Sustainability and Innovation, https://doi.org/10.1007/978-3-031-08313-6_8

1 Introduction

In facing current climate and ecological crises, substantial hopes and expectations are placed on technological solutions, what traditionally has been defined as the ecomodernist approach to the environmental transition and an important basis of the green growth paradigm (The World Bank, 2012). Green growth relies on innovation (Bowen & Hepburn, 2014) of new technology not least for a low-carbon development but also for reduction of toxic pollutants, recycling of nutrients, and savings of scarce resources. While green growth is actively promoted by OECD and other international organizations (GGGI, 2014; Grunwald, 2018), it has been exposed to criticism both in terms of its practical feasibility (Hickel & Kallis, 2019; Ward et al., 2016) and its implications (O'Neill, 2020).

It is well established that developing new technologies[1] is not a straightforward, simple, or linear process and no universal recipe exists for bringing them successfully to the market. The idea of technology innovation as a complex, iterative process was pioneered with the theory of creative destruction by Joseph Schumpeter (Ziemnowicz, 2013). According to Dosi (1982), "there is a complex structure of feed-backs between the economic environment and the directions of technological changes": in early stages, social and economic conditions shape the direction of technological innovation, technology development then gains momentum on its own, and technologies are produced, but again social and economic conditions determine ex post their survival, for example, via market selection. According to the early work of Grubb (2004), a key issue of green technology development is that it occurs within a field where policies are uncertain and market incentives are scarce. Grubb identifies six stages of technology innovation from an innovation chain perspective (basic R & D, technology R & D, market demonstration, commercialization, market accumulation, diffusion) and three stages from a public policy perspective (where each stage requires a larger amount of funding and displays a gradual shift from public to private funding). Grubb questions the conventional dichotomy between supply-pull and demand-push policies for innovation and argues that while supply-pull policies affect the early stages, demand-push policies affect the later stages. In between, we find the infamous "valley of death" (Murphy & Edwards, 2003) where the crucial challenge is to obtain the support of private investors.

Jacobsson and Bergek (2011) build on the technological innovation system framework (Hekkert et al., 2007) to stress the importance of a systems thinking approach while identifying weaknesses in processes of environmental innovation,

[1] The term "green technology" is used here to refer to a technology that is considered environmentally friendly based on its production process or supply chain. We chose this term to match the "green growth" terminology and among various synonymous terms such as "environmental technology," "green tech" and "cleantech" (Investopedia, 2019), and "environment-related technologies" (Haščič & Migotto, 2015).

especially in emerging fields, and with the purpose of defining policy to support such processes.

In their literature review-based study, Bergek and Berggren (2014) stress how "Innovation is not a linear progression from R & D to market introduction, but a complex, interactive process" and define innovation as the "development, industrialization, and commercialization of technologies" and in particular environmental innovation as "the development, market introduction and early diffusion of new or refined technologies which reduce undesirable emissions." They identify experiments, early applications, niche markets, as well as research, development, design, production, and marketing as key elements, or steps, in the innovation process. They highlight that in the innovation process there are several loops between these stages, and therefore the process is not linear.

A comprehensive review of the drivers of environmental innovation is provided by Hojnik and Ruzzier (2016). While Grubb (2004) and Bergek and Berggren (2014) acknowledge but refrain from explaining the complexity of the environmental innovation process, Hojnik and Ruzzier (2016) embrace fully this complexity and provide a very extensive list of possible drivers and stages of this process. The resulting picture is rich but lacks a clear structure (e.g., R & D is considered both a stage and a driver) and appears unfocused.

The studies mentioned above have addressed either environmental innovation in general or specific sectors, like the automotive industry. An important issue remains how to define what are innovative technologies relating to environmental protection. One proxy indicator for innovation is the patenting of technologies, as a patent can be seen as an indicator of the novelty of an invention. Patents are classified systematically with the Cooperative Patent Classification (CPC) that allows for identifying comparable technologies and inventions with environmental relevance (Haščič & Migotto, 2015). Numerous studies take patents as their data basis using quantitative statistical approaches to study innovation (Dolfsma & Leydesdorff, 2011; Johnstone et al., 2010; Palage et al., 2019) and push/pull policies (Cantner et al., 2016; Costantini et al., 2015). Yet, patent databases also provide information on the actual patent inventors. Breschi (2000) used inventors' addresses to study the regionalization of environmental innovation. Karvonen et al. (2014) and more recently Scandura (2019) carried out interviews with inventors that have been patenting their technologies and the reasons for patenting.

The Nordic region represents a promising case of environmental innovation due to several characteristics. Nordic countries have rather ambitious policies for environmental protection and low-carbon development (Hildingsson et al., 2019; Westholm & Beland Lindahl, 2012), and they share a long tradition for building a knowledge-based society with substantial public financial support for research and development in general and for green technologies in particular. Specific cases like the Danish wind turbine industry are well-known examples of successful environmental innovation (Mazzucato, 2013), reflecting a wider legacy comprising also district heating, insulation materials, hydro, smart grids, and other (Andersen, 2020). Sovacool (2017) provides a comprehensive assessment of comparable traits and differences in building a common environmental narrative for each of the five Nordic countries. A critical conclusion reached by Sovacool (2017) is that technological breakthroughs are frequently obtained through close collaborations between business sector and public research organizations (Nielsen & Pedersen, 2013). Despite the high level of Nordic environmental innovation, new green technologies might not succeed in upscaling, suffering low acceptance by industry and consumers, without major changes in public policies, institutionalized supply chains, and other surrounding conditions. Illustrative in this respect is the case study by Fevolden and Klitkou (2017) regarding the failure to upscale technological innovations in the biofuel domain in Norway. The question at this point is what we can learn from a closer analysis of the technology-focused green growth efforts ventured in the Nordic countries.

The aim of this study is hence to provide a broader empirical analysis of which key factors shape the journey of new green technologies from their early conception to their large-scale adoption. In accordance with previous research, this journey is understood to coincide with the process of "environmental innovation" (Bergek & Berggren, 2014). This study contributes with new knowledge to further unravel the complexity and the challenges in the process of environmental innovation. We analyze and approach this process under a critical lens that questions the simplistic techno-optimist narrative that innovation in technology can be a straightforward solution to climate change and other environmental issues. Our expectation is that the main implication of this research will be an improved understanding of the possibilities of technology-driven green growth and that the findings will allow inventors, researchers, and policymakers in this domain to improve the way they carry on, investigate, and regulate green technology development. The research can be defined as "explorative qualitative" and combines case study and historical and narrative research designs: informed by prior knowledge from literature and focusing on deriving a common narrative from the direct experiences and storylines of informants. The scope of the analysis is with patented technologies in the environmental domain of the Nordic countries.

2 Methods

2.1 Analysis of Environmental Patents in the Nordic Countries

From the patent database of the European Patent Office (EPO, 2019), we retrieved data for all the patent applications submitted in the period 2000–2017 in the five Nordic countries: Denmark, Finland, Iceland, Norway, and Sweden. To restrict the search to technologies with an environmental relevance, we used the list of Cooperative Patent Classification (CPC) codes identified by a recent OECD report on environmental innovation (Haščič & Migotto, 2015). After retrieving the relevant bibliographic information on these patents (application number, CPC codes,

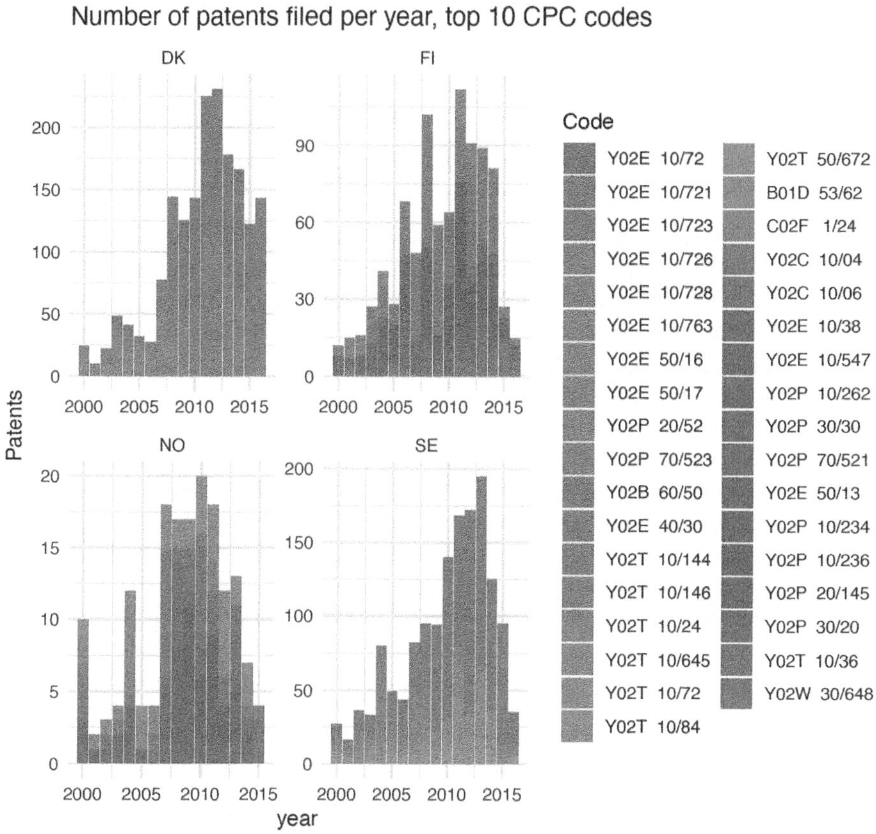

Fig. 1 Number of green patents filed in Denmark (DK), Finland (FI), Norway (NO), and Sweden (SE) in the period 2000–2017. The top ten Cooperative Patent Classification (CPC) codes are included in each country, ranked based on occurrence (number of patents classified according to each code). The description of each CPC code can be accessed at the website of the European Patent Office: https://worldwide.espacenet.com/classification

inventors and their affiliation, title, country, year of submission, etc.), we screened the dataset to identify the national patterns of environmental innovation. Since name and affiliation of patent inventors are available from the database, we used this information to identify and interview a subsample of inventors for an in-depth analysis of the challenges experienced during green technology development toward upscaling and commercialization. Figure 1 shows the number of "green patents" filed per country, for the top ten CPC codes in each country ranked based on occurrence (number of patents classified according to each code). The figure shows a common trend across Nordic countries: the number of green patents has been increasing steadily since 2000 (the decrease after 2015 is attributable to delays in reporting). Looking at the colors, the figure also indicates that each of the Nordic countries displays a different pattern in terms of type of technologies patented suggesting country-specific strongholds or preferred areas of green technology development.

2.2 Interviews with Inventors of Green Technologies

To define a manageable sample of inventors to approach, we selected patents classified according to the most frequently occurring CPCs in each of the Nordic countries under analysis. Specifically, these relate to energy efficiency in buildings based on ICT, energy generation from nonfossil fuels, electrification of vehicles, and general improvement of production processes causing GHG emissions. However, we omitted green technologies related to renewable energy, given that innovation in the, for example, wind and hydro domains has been studied extensively already, having achieved a breakthrough (Mazzucato, 2013). We chose to perform semi-structured interviews as they allowed us to retrieve a detailed storyline behind each patent and technology and to better describe the process and factors shaping it. The interviews were conducted in the period between September 2018 and March 2019, recorded, and transcribed. Supporting Information includes full details on the interview procedure (SI.1) as well as the interview guide (SI.2).

2.3 Qualitative Content Analysis of Interview Data

The interview data were subjected to qualitative content analysis to identify recurring *codes* and common *themes*. The method used can be defined as a "thematic analysis" (Braun & Clarke, 2006) or a "content analysis" (Erlingsson & Brysiewicz, 2017), although the term "qualitative content analysis" is preferred here to clarify that no quantitative analysis was performed (Vaismoradi et al., 2013). The analysis proceeded iteratively in different coding rounds, until a satisfactory combination of codes and themes was found. The coding was deemed satisfactory when it fulfilled two conditions: a number of recurring and nonoverlapping codes and themes could be identified, and the coding allowed to maintain, explain, and reconstruct the narrative behind each interview. The set of codes and themes allows us to define a matrix specifying the topics covered by each code/theme interface, which we define here as our *data structure*—using the terminology of Gioia et al. (2013). Each interview was divided into quotes, and the main topics addressed by each quote were identified (a quote could address several topics) and summarized into a *condensed meaning*. Finally, all the quotes and condensed meanings were organized into a database, and those belonging to each code were filtered out and analyzed together to derive a summary. An example of coding and condensed meanings is provided as Supporting Information (SI.5).

Table 1 Number of green patents per Cooperative Patent Classification (CPC) code per country. The top 20 codes are included in each country, ranked based on occurrence (number of patents classified according to each code)

Country	CPC code (Number of patents) Sub domain
Denmark	Y02E 50/17 (411) Biofuels
	Y02E 50/16 (297) Biofuels
	Y02P 70/56 (144) Manufacturing of batteries and fuel cells
	Y02P 30/20 (134) Bio-feedstock
Sweden	Y02B 60/50 (1811) Information and communication technologies
	Y02T 10/7005 (385) Electric vehicles
	Y02T 10/645 (214) Electric vehicles
	Y02T 90/14 (201) Electric vehicle charging
	Y02T 10/7072 (183) Electric vehicles
	Y02T 10/705 (181) Electric vehicles
	Y02T 10/7077 (176) Electric vehicles
	Y02T 10/7275 (174) Electric vehicles
	Y02T 10/72 (153) Electric vehicles
Norway	B01D 53/62 (118) Emissions abatement from stationary sources
	Y02E 50/16 (55) Biofuels
	Y02P 20/152 (54) General improvement of prod. processes causing GHG emissions
	C02F 1/24 (50) Water and wastewater treatment
Finland	Y02B 60/50 (1275) Information and communication technologies
	Y02P 30/20 (414) Bio-feedstock
	Y02E 50/13 (179) Biofuels
	Y02P 20/145 (165) General improvement of production processes causing green-house gas (GHG) emissions
	Y02E 50/14 (152) Biofuels
	Y02E 50/16 (99) Biofuels

3 Results and Findings

3.1 Survey of Inventors

We retrieved data for more than 11,000 patents. After filtering the patents for the top 20 environmentally related CPC domains per country,[2] the sample included 2150 unique inventor names. Additional details on the sample of patents used for the selection of inventors are provided in Table 1 and further expanded in Supporting Information (SI.3) with the full description of each code.

It was relatively straightforward to approach inventors affiliated to public organizations and universities, or to SMEs, as their contact email was usually available on the website of their organization. For patents with several inventors, priority was given to the first inventor in the list. A total of 134 invitations was sent via email, and 41 inventors accepted to be interviewed (12 Danish, 12 Swedish, 7 Norwegian, and 10 Finnish). Some of the respondents are listed as inventors for more than one patent, and although each interview focuses on a specific technology, inventors often were

[2]Iceland was excluded due to its insignificant number of green tech patents.

able to draw parallels with other inventions or closely related patents of theirs, so the total number of patents covered in this analysis is 129. The inventors interviewed included senior scientists or professors at universities, senior staff in management positions at large private companies, and CEOs or CTOs of small-medium enterprises or start-ups. Still, in some instances, it was not possible to retrieve the email or contact, even if a profile page was found on social media platforms (LinkedIn, ResearchGate) or a company website. Additional details on the country, invention area, and affiliation (industry/company or research/academia) of the inventors are provided in Supporting Information (SI.4).

3.2 Understanding Environmental Innovation in the Nordics

Figure 2 provides a graphical representation of how four main factors interact and affect the different stages of environmental innovation. This framework was derived from our analysis of the interview data while informed by existing conceptualizations of innovation (see above).

The four identified factors are *organization*, *technology*, *market*, and *funding*. The *organization*[3] factor explains how the social and cultural environment and the entity where the inventor belongs or is associated with affect the technology development process. The *technology* factor covers all technical aspects that are specific to the environmental invention under analysis. The *funding* factor covers all financial aspects both in early and in later stages of technology development. The *market* factor explains how political and economic boundary conditions affect the technology development process. The interfaces between the four factors allow us to disaggregate the life cycle of green technologies into seven different stages (Cf. Fig. 2): *R & D*, *start-up*, *patenting*, *upscaling*, *dealing with risk*, *commercialization*, and *adoption*. The stages are not necessarily consecutive, and each of them is affected by at least two different factors. The first stage is frequently the research and development (*R & D*), which is strongly shaped by the organization where it is done and by the funding that makes it possible. *Patenting* is done typically (but not necessarily) after *R & D* and is strongly dependent on the organizational culture and on the characteristics and type of technology being developed, for example, its readiness level. *Start-up* differs from *R & D* as it requires different organizational settings and funding (e.g., the creation of a separate company and securing seed funding) and because in this stage there is a strong focus in highlighting the innovative and environmental potential of a technology. The *upscaling* stage roughly corresponds to what other literature sources and some of the inventors interviewed define as "valley of death," a situation where financial and technical challenges are substantial and more determining than organizational and market

[3] A fitting name could also be "milieu," intended as "the people, physical, and social conditions and events that provide the environment in which someone acts or lives" (Cambridge dictionary).

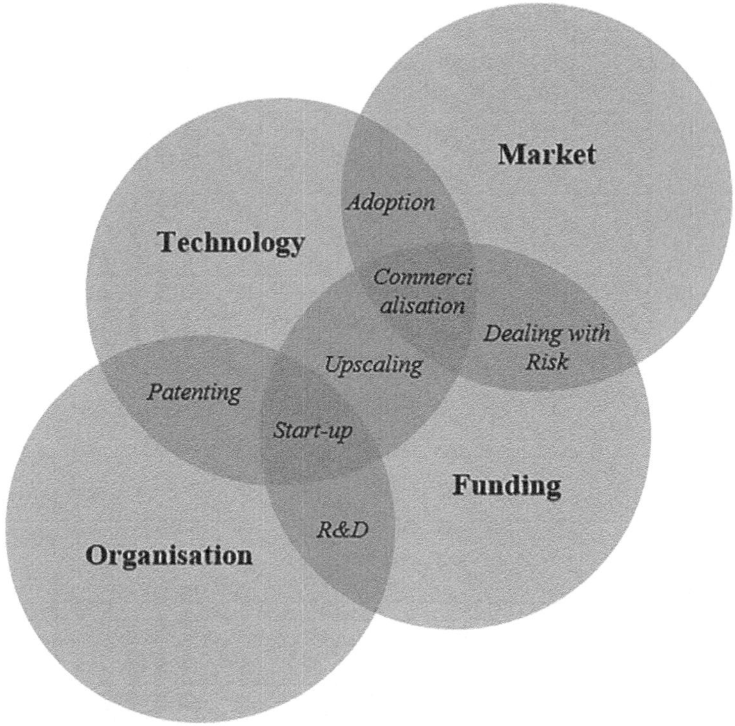

Fig. 2 Inductively derived structure fitting the interview data on environmental innovation in the Nordics. The four main factors, organization, technology, funding, and market, influence the technology life cycle that occurs in seven different stages: R & D, start-up, patenting, upscaling, dealing with risk, commercialization, and adoption

factors. The funding from commercial partners, market prices, and competitors and technology-specific characteristics are all shaping the *commercialization* stage according to our interviews. The *adoption* of the green technology is then influenced by the technological factors such as the receiving infrastructure and the target market segment. Finally, a recurring issue is *dealing with risk*; in this stage, the most important factors are market-related like the role of existing policies and regulations and funding-related such as how investors perceive the market.

To allocate the observations and issues from the content analysis of our sample of green patents to the matrix presented in Table 2, each *code* corresponds to a *factor*, and each *theme* corresponds to a *stage*. Thus, only the terms *factor* and *stage* are used in what follows. For example, the interface between the factor *funding* and the stage *R & D* covers all information on the role of funding during the research and development on the green technology under analysis. The reason why some of the factors do not intersect some stages is because they did not appear in the interviews or appeared only to a very limited extent. For example, when explaining about the *patenting* process, the inventors focused to the largest extent on the reasons for

Table 2 Matrix of the topics covered in each factor/stage interface and summary of findings. These are also the criteria for classifying a quote according to a specific topic. In bold, the common topic addressed at each stage/factor intersection

Stage factor	R & D	Start-up	Patenting	Upscaling	Dealing with risk	Commercialization	Adoption
Organization	**Organization where the R & D activities were carried out** Private-public collaboration: testing, problem-solving, or externalize the risky ideas via partnership in research project or work for hire. In-house project at university (explorative) or company (clear business-oriented application)	**Role of organizational and cultural capacity** Starting up company optimal setting to access further funding as well as forced choice to be able to patent a technology or continue the development. Need to build on existing knowledge, possess the right mentality and know-how	**The reasons for patenting** Expensive and complicated process necessary to increase success chances. Restricting and protecting the (non)use of the tech. Patents as "business cards" to attract investors, build a reputation, reach goals set by tech transfer office				
Tech		**How the technology is environmental and innovative** Bio-based techs: improve feedstock processing efficiency, use complex feedstocks, new	**The readiness level of patent** Tech far from commercialization when the patenting starts. Time needed for environmental innovation underestimated. Systems of patents	**The technical challenges in upscaling** Improve process efficiency by reducing time and cost of production. Both tech-neutral and tech-specific.		**How the commercial level is reached in technical terms** Green techs expected to operate on massive scale, but the expected commercial form is tech-	**The receiving infrastructure and industry** Techs targeting a receiving industry sector and are easily integrated in receiving facilities versus techs

		bioproducts. Techs for decentralized energy production and storage: faster mass production techniques, more efficient materials/design. Other techs: reduce toxic emissions and nutrient loss	building on top of each other in portfolios to showcase to investors	Upscaling takes long time and has high uncertainty. Technological challenges not as difficult as financial ones		specific: large feedstock processing plants, mass-produced high-tech components, medium-scale decentralized biorefineries	that target or cross multiple industry sectors and for which the receiving infrastructure is not yet mature
Funding	**How the R & D was funded** In early stages, either public research grants or private internal company funds. Early funding essential for R & D and well available but does not necessarily help moving into successive stages	**Securing seed funding** Useful to apply for patents, put together a team, networking activities, start a business. Shift from research to market value focus. Developing the people around the tech. Incubators and private investors key role. Lack of public funding		**The financial challenges in upscaling** Requires exponentially larger investments than previous stages. "Valley of death": lack of large public funding and of private investors. "Catch 22" problem: investors want to see the tech work at scale before investing but	**How investors perceive the market** Risk-averse and interested in stable long-term market conditions, rarely achieved. Expect that new regulations, policies, and taxes will be introduced. Interested in mature technologies and fast returns, while green techs show	**The role of commercial partners** Important to license the tech and cover production costs, bring specific know-how, facilitate integration, or further test it in real-world settings	

(continued)

Table 2 (continued)

Stage factor	R & D	Start-up	Patenting	Upscaling	Dealing with risk	Commercialization	Adoption
				investment required to proof that	slow growth and long payback times		
Market					**The role of policy and regulation** Pull policies as de-risking instrument (taxation of fossil fuels, incentives for renewables) and policies creating niche markets foster the innovation. Other policies (safety, GMO, food safety, waste) can obstacle it in indirect ways. Some policies concurrently favor and disfavor it (biofuels)	**The role of prices and competitors** Competing with incumbent refined technologies, other emerging green techs, and identical techs by other inventors. Large companies adopt hard strategies to exclude start-ups and SMEs. Low prices of fossil resource major obstacles, incentives can create specific lock-ins	**The target market segment** Different strategies: targeting multiple target markets and ensuring flexibility versus developing green techs that target specific niche markets around the world, very often outside the Nordic region, versus creating new markets

patenting and the readiness level of patent, and these two topics can be referred to the *organization* and *technology* factors, respectively, whereas considerations about funding of patents and receiving market conditions were very limited or absent.

3.3 Findings of the Qualitative Content Analysis

In the following sections, we report our findings and focus on describing cause-effect mechanisms: how each factor influences different stages of the environmental innovation process. Each section proposes a synthesis of all similar outcomes (quotes belonging to the same stage/factor interface) from all interviews. These findings are summarized in Table 2. Selected quotes supporting the statements that we make in the following sections are provided as Supporting Information (SI.6).

3.3.1 How the Organization and Social Environment of Inventors Affect Green Technology Development

The role of the organization where inventors belong and in broader terms of their social environment is particularly important in the stages of R & D, start-up, and patenting.

Organization and R & D

We observe that in the R & D stage, often the collaboration between private and public organizations is a strong positive factor and even a necessity to initiate and move the technology development forward. In the Nordic region subject to analysis, companies collaborate with universities for different reasons, for example, the need for specific testing, to solve specific development problems or even as an explicit strategy to externalize the risky research ideas. The collaboration conditions vary from actual partnership in a research project where private and public organizations share goals and plans to situations of work for hire where the university staff is bought out for specific short-term tasks. It is interesting to note that even in a collaboration, the two types of organizations might have different goals and motives or roles, so the technology development takes different directions depending on the context where it is conducted. We also observe cases where the research and development were conducted fully within a university, usually within an explorative research project, or fully within a company triggered by a specific need and within a clear business-oriented application. Summing up, our data show that the support of an organization and especially the collaboration between organizations are necessary factors shaping environmental innovation in the R & D stage.

Organization and Start-Up

In the start-up stage, three main organizational mechanisms can be identified: starting up a company to formally enter the innovation sphere, the overall positive support given by national and international collaboration networks and other country-specific organizational conditions, and the importance of building on existing knowledge and of possessing the right mentality and know-how.

We observe that inventors are very often forced to change the formal setting of their activities to pursue environmental innovation, for example, shifting from being a university employee to the funder of a start-up company. This is due to different reasons, for example, because the university or the large company where the inventor is employed is not interested in supporting the patenting process or further research in the technology or even sees the technology as a direct competitor or as not aligned with its objectives. Creating a start-up company is often a prerequisite to file patents (particularly in Sweden) and a positive factor as it can allow to access new types of funding to continue the innovation process, for example, public funding from innovation funds. The "start-up company" appears thus to be an appropriate organizational setting to pursue a business idea and enter the network of incubators, venture capital investors, and innovation funds. The shift from university or large private company to start-up company is perceived as a suffered choice, as it involves a substantial workload and initiative-taking attitude. The shift might not even materialize as the risks are deemed excessive or the perspectives of leaving a secure position discouraging.

Organization and Patenting

In general, patenting of green technologies is perceived as an expensive and complicated process that is nevertheless done to increase the success chances of the technology either because it allows for restricting and protecting the use (or nonuse) of the technology or because patents are used as "business cards" to attract investors. However, there are factors related to the organizational settings and culture that affect the patenting process in several ways. In Sweden, making a start-up is a prerequisite for academic inventors to file and exploit patents. Different attitudes in the organization's approach to patenting influence the number and quality of patents produced also in private companies. The reasons for patenting might be completely unrelated to the quality of the patent and are rooted in other drivers that are organization-specific, for example, the need to build a reputation or to reach goals set by the technology transfer office. Several inventors state that the technology transfer office of the university played an important but not always positive role in determining the development of new green technologies. Summing up, the patenting stage is affected by organizational and cultural factors both positively and negatively for green technologies developed in the Nordic region.

3.3.2 The "Technology Factor" in the Life Cycle of Green Technologies

Technology and Start-Up

We grouped here all quotes describing how the technology under analysis is environmental and innovative, and we also use this category to describe the technology types under analysis. The first group is technologies for the processing of biomass resources into biofuels, biogas, and biopolymers. The innovative aspect lies in either making feedstock processing more efficiently or using different and more complex feedstocks or residual ones, or producing new bioproducts from existing feedstocks, and in general avoiding the use of fossil fuels. The second group is technologies for decentralized energy production and storage, in particular fuel cells. Here, the innovation lies in new manufacturing techniques toward mass production and in efficiency improvements via the use of new materials and design. The third group is more mixed and includes technologies for production of fertilizers, sanitation, and enzymes; these can be innovative as these reduce toxic emissions and nutrient loss.

Technology and Patenting

The readiness level and value of a patent and its relationship with other patents are technology-specific. Most inventors stated that the technology was far from commercialization when the patenting process started or at least not as close as anticipated. Several inventors admitted having underestimated the substantial time needed for environmental innovation. We observe several cases where environmental patents are part of a "system of patents," where one or more aspects around the same technology, or different smaller technologies in a bigger process involving all of them, are patented together. In this way, the patents "build on top of each other" and allow the inventors to build a "patent portfolio" to showcase to potential investors. Partly because of this systematic approach to patenting, inventors admit that not all inventions possess the same novelty and value and identify "incremental," "conceptual," "introductory," "method," and "product" patents, showing how patent portfolios can be highly diversified in type and quality depending on the technology. However, we found examples of patents filed for technologies that were very close to commercialization, mainly because of combining existing technologies in a different way.

Technology and Upscaling

We differentiate between technical challenges and financial challenges in upscaling (cf. next section). We identify challenges that are technology-neutral, such as improving process efficiency by reducing the time and cost of production, and a general agreement across inventors that the upscaling stage takes long time and is characterized by substantial uncertainty. Another important general issue is the need for specific technical knowledge in this stage and for having a good interdisciplinary team. Then, we identify technology-specific challenges. For technologies in the bioeconomy area, upscaling means increasing size, for example, building and operating a large plant with bigger machinery, shifting from batch to continuous processing, and processing a large amount of feedstock; dealing with the natural variability in feedstock, materials, and processing conditions (e.g., working at high temperature and stable conditions), and the integration of the new technology with other new and old ones in a process or system, in order to exploit synergies such as heat recovery. For technologies like fuel cells instead, the technical challenge is to refine the manufacturing process to improve the product quality (lifetime, reliability) and to mass-produce a high-tech component by automating highly detailed manual work. Fuel cell inventors report the need for extensive product testing under different conditions, at scale, and close to customer. An important result, which we highlight, is that several inventors found the technical challenges not as difficult to overcome as the financial ones.

Technology and Commercialization
How the commercial level is reached in technical terms is also technology-specific. A key feature of green technologies, according to our analysis, is that they are expected to operate on a massive scale in real life. For example, biomass processing technologies are expected to be implemented in several very large plants to process huge amounts of feedstock, for example, to produce biofuels. Conversely, fuel cells are relatively small, high-tech components that are expected to be mass-produced in large factories using robots with high degree of automation. For green technologies related to enzymes and fertilizers, the expectation is to produce massive volumes of products in large-scale facilities. Some of the technologies of our study have been developed for commercial operation in small- or medium-scale facilities (tank- or container-size), which should allow for decentralized biorefinery or nutrient production close to the feedstock supply and reduce the upscaling challenges. Thus, what commercial scale means in practice depends on the technology, and therefore different technologies face different challenges in reaching such scale, and it is not possible to generalize.

Technology and Adoption
Finally, we highlight here the role of the receiving infrastructure and industry in facilitating or impeding the adoption of green technologies. Based on our responses, we can make a distinction between two cases: technologies that are developed aiming for a receiving industry sector (e.g., pulp and paper industry, petrochemical industry, energy production facilities, wastewater treatment plans, etc.) and that can be easily integrated in receiving facilities, at least in theory, and then green technologies that target multiple industry sectors or cross different sectors and for which the receiving infrastructure is not yet in place or not mature, thus impeding their adoption. This mirrors the difference between incremental and radical innovations (Beck et al., 2016) for the case of green technologies.

3.3.3 The Role of Funding in the Development, Upscaling, and Commercialization of Green Technologies

Not surprisingly, funding is a key factor in several stages: R & D, start-up, upscaling, commercialization, and dealing with risk.

Funding and R & D
Essentially, all the inventors in our survey secured some type of funding in the early stages of the technology development, being either public funding from research grants (typically the case of universities and SMEs) or private funding internal to a large company. We acknowledge that there might be a certain selection bias (only the technologies that get initial funding are patented) but the obvious conclusion is

that the availability of funding is essential for R & D and that in the Nordic region this type of funding is to a large extent available. However, the availability of public funding sparks substantial R & D but does not necessarily help the technologies moving to the next stages.

Funding and Start-Up
More critical is the effect of securing seed funding in the stage of start-up. Seed funding is used to apply for patents, to put together a team, to perform networking activities, and to start a business. Seed funding has therefore a role in facilitating the progress of the invention from a purely research item to an item with market value; it's for developing the people around the technology rather than the technology itself. Incubators and private investors play then the key roles to facilitate or unlock the access to seed funding. The inventors interviewed highlight a lack of public funding in this stage. Several respondents stress the importance of soft skills in getting seed funding (and also funding for upscaling; see later) and having "the right guy" or "the guy who was good at getting money" as part of the team. Again, the different funding formats shape the environmental innovation process by transforming the team working on the technology from a technical team to a business-oriented team, by including persons with softs skills, networking, marketing, and business planning, and to strengthen the business idea around the technology rather than the technology itself.

Funding and Upscaling
The *upscaling* stage is characterized by substantial financial challenges, because the size of investment required is large and exponentially higher than in the *R & D* and *start-up* stages and the sources of capital are scarce. The large investment is needed in part because a larger team needs to be maintained, and the technological development takes long time and almost always more than anticipated by the inventors, but mainly because there is a need to produce in massive amounts. This means either building a large plant, as, for example, in the case of biomass processing technologies, or mass-producing a series of highly technical components at a low price, as, for example, in the case of fuel cells. The development work is labor- and infrastructure-intensive, which needs large funding in the size of millions of euro. The lack of funding in this stage is what inventors refer to as the "valley of death" of green technologies. Among the challenges that we identified is the lack of, or scarcity of, private investors, these defined as private investors interested specifically in green technologies. One interesting aspect is what inventors identify as the "catch 22" problem or "chicken-and-egg" problem. In short, in this phase, investors want to see the technology at work in real scale before investing money on it, but the funding is required precisely to proof that the technology can work at large scale, because there are substantial differences between small- and large-scale productions (cf. previous section on technology). One solution, mentioned for companies facing the challenge of mass production, is to subsidize the costs internally to get the market created in the first place and therefore sell the product under price to make it viable for customers. Most importantly, what is missing in the *upscaling* stage compared to the *R & D* and *start-up* stages is public funding, as there are no public grants

available of the size needed, and the government is reluctant to make such large-sized investments because it would be breaking the principles of market competition, favoring one company rather than others. For demonstration projects, EU funding is seen as a valuable option for many inventors, although the high co-funding requirements, the low success rates, and the red tape discourage applicants and represent an obstacle for SMEs to join these projects. While some inventors see more public funding as the solution to the valley of death problem, others argue that governments instead need to have a more ambitious vision and to create long-term conditions and niche markets, which will give security and attract investors. This mirrors the "pull-versus-push" debate that we know from the literature. It is interesting to observe that some respondents declared to have experienced no financial challenges at all. These are inventors working within large private companies, which have developed technologies in house, taking a substantial risk but also heavily supporting the technology with large capital investments.

Funding and Commercialization
At the stage of *commercialization*, external partners play a major role in terms of funding. At this point, the technology is technically and economically viable to a sufficient extent that a market player can take it up. Thus, the availability of commercial partners contributing with industrial and financial capital to the implementation of the technology becomes key. Commercial partners are important financially as they can license the technology and cover the costs of production but also from a development perspective as they bring specific know-how, for example, in mass production, or contribute with the integration of the technology with other industrial processes or further test it in real-world settings. These are aspects that can further shape the technology and increase its market value. Many of the green technologies in our samples were developed in collaboration with an external partner from the very inception; this is the case for fuel cells and biomass processing technologies. In some cases, this turned out to be advantageous because it ensured a smooth transition to the commercialization stage. Yet it was counterproductive in cases where the commercial partner decided to suddenly pull out from the operations. Since the development highly was partner-driven and targeted to the specific needs of the partner, it arrested abruptly.

Funding and Dealing with Risk
The interface between the factors *funding* and *market* at the stage of *dealing with risk* is here interpreted as the way the investors perceive the market. Our respondents are generally agreeing that investors in green technology are risk-averse and keen to ensure stable market conditions in the long term, which is only rarely achieved (cf. next section on market). This expectation of a secure market is strongly related with the expectation of new regulations, policies, and taxes which will be introduced. Investors are interested in mature technologies at high TRL and fast returns, while green technologies have slow growth and long payback times, features that do not exactly underpin investment. Interestingly, although the green technologies often have evident societal and environmental value, the focus of investors is on obtaining economic returns rather than on sustainability as such, and they do not distinguish

between green and nongreen technologies—according to the inventors we interviewed.

3.3.4 How Market Conditions Shape the Path of Green Technologies Toward Commercialization

Market conditions become a relevant factor to the stages of *commercialization*, *adoption*, and *dealing with risk*.

Market and Commercialization

In the *commercialization* stage, what matters is the key role of market prices and competitors. Green technologies are competing with incumbent, older technologies already in the market, which have been perfectionated and refined over the years and are thus at a much higher development stage. Moreover, they are competing with other types of emerging green technologies developed for the same purpose or market segment (e.g., biofuel production or transport) or with identical technologies developed by other research groups, as there is a substantial environmental R & D in the Nordic region. This accentuates competition for the financing and commercial partners to test the technology and produce it. Large companies are tough competitors for start-ups and SMEs, not only because of their financial strength but also because they can adopt strategies like licensing the new technology and then deliberately keep it out of the market or pursue a parallel development of secure technologies in-house and risky ones externally. According to our respondents, the low prices of fossil resources are a major obstacle for developing many technologies in the energy and transport sectors (biofuels, fuel cells). But this is true in other cases of competing green technologies too, as, for example, the public support for biofuel prices is favorable to developers of technologies in the biomass-to-fuel pathway, causing lock-ins to the development of technologies for upgrading biomass, in the biomass-to-chemical pathway (e.g., the production of biopolymers).

Market and Adoption

We observe that the type of target market segment for the technology can affect its success in three different ways, according to our data. One strategy is to develop green technologies aiming for multiple target markets and ensuring flexibility. For example, a technology for biomass processing can be used to target the transport market or the feed market or the transport market and the pharmaceutical one. A second trend is developing green technologies that target specific niche markets around the world, very often outside the Nordic region, for example, in Asia due to better availability of resources or need for decentralized energy production or in the United States because of higher public support for biofuels. The third option is to create a market that is not yet there, for example, by crossing traditional supply chains or creating new services.

Market and Dealing with Risk
Finally, in the interface between *market* condition factor and the *dealing with risk* stage, we placed the role of policy and regulation. We identify several mechanisms for how policies and regulation favor or disfavor environmental innovation in the Nordics by increasing or decreasing the risks in the process, respectively. We confirm previous research on this topic (Costantini et al., 2015; Mazzucato, 2013; Nemet, 2009) and in particular the strong effect of pull policies as de-risking instrument, and we provide here some examples. Inventors can clearly acknowledge the positive effect of taxation of fossil fuels and incentives for renewables (such as blending mandates for biofuels and incentives for biogas) and policies creating niche markets have on fostering the technological development and facilitating the investments. There is a well-known dilemma between technology-neutral policies and "picking the winners" policies that create niche markets for new technologies, if these are truly innovative. We observe that certain policies and regulations not directly targeting environmental innovations can exercise influence, sometimes in bizarre indirect ways, for example, EU regulations on GMO can impede the development of specific biotechnologies with environmental applications; fire and safety regulations can slow the upscaling of biomass processing technologies; regulations on food safety can impede the use of captured CO_2 in the food industry (beverages); regulation on plastic waste disposal can favor the development of bioplastic production technologies. Finally, policies may concurrently favor and disfavor environmental innovation, for example, in the case of incentives for biofuel production which favor technologies to produce low-value products from biomass feedstock and disfavor technologies to produce high-value products (biopolymers) from the same biomass.

4 Discussion

In this section, we consider the robustness and validity of the results, as well as their novelty in the light of previous studies.

Robustness of Data and Approach and General Validity of Results
First, we focus on the limits due to sample and data. Patents are not necessarily synonymous with innovation, and this is also confirmed by many respondents. The use of patent databases as data source to sample green technologies, and then identify inventors, was particularly useful in this study. It allowed access to in-depth information and to reach out to many potential respondents. The drawback of this approach is that technologies that are in the making and for which no patent application has been filed are excluded from the analysis, as well as organizational innovations. The sample composition is balanced across the four Nordic countries but is skewed in terms of age, gender and ethnicity, and affiliation. Nevertheless, we believe this reflects the fact that actors in the Nordic green innovation are predominantly white, male, highly educated professionals in academia and industry. It was

beyond the scope of this analysis to investigate how age, gender, or ethnicity of inventors affects the technology innovation process (Kronsell, 2013). The advantage is that since most of our respondents occupy high-level positions in their organization and have considerable experience (several had published numerous patents), they provide highly knowledgeable informants, underpinning the reliability of their statements. We flag that we found it easier to approach inventors in academia and in small companies, rather than in industry, as their email address is publicly available, and they showed a higher level of engagement. Nevertheless, our sample is equally balanced between inventors in academia and outside it. Finally, there are limits due to the technologies under analysis, with two dominant groups of technologies in the bioeconomy and fuel cell area and a third mixed group of technologies (fertilizers, enzymes, CO_2 capture) that should be considered when assessing the general validity of these results.

We may speculate whether these findings are valid beyond the Nordic region and beyond the green technologies covered by our survey. Our assessment is that some finding might have general validity in Europe and high-income economies, for example, the role of policies and market forces to influence the green technologies under analysis works with other green technologies too, as confirmed by several previous studies for the cases of renewable energy in Germany and (Cantner et al., 2016; Peters et al., 2012) and electric vehicles in 15 OECD countries (Choi, 2018). Instead, drivers of innovation might differ between green and other technologies (Cuerva et al., 2014), and the way organizational settings affect the process might be more specific of Nordic countries and their collaboration models to be valid for other geographies. Our interviews highlight that a key feature of green technologies is that they often are aimed at performing on a large scale, requiring vast infrastructure. This is fundamentally different from other types of technologies like information technologies that have more exponential growth compared to the environmental ones (slow growth). Moreover and building on a quote from our interviews, "There is a narrative that IT tech leads to faster returns. Investors as well prefer the IT, they want quick-in and quick-out. Instead, the green techs are driven from ethical concerns about sustainability." We suggest that the development and diffusion of green technologies are motivated partly by factors that go beyond the pursuit of economic returns, and our results might therefore not be applicable directly to other types of technologies.

Similarities with Previous Studies

A similar inductive approach of coding a dataset to see the emergence of a theme was employed by Nemet et al. (2018): based on the empirical analysis of 511 demonstration projects in 9 low-carbon technology areas—from solar thermal electricity to cellulosic biofuels—they highlight "valley of death" challenges to green technologies and propose solutions to overcome these challenges.

The structure proposed in Fig. 2 shares similarities with the models of Grubb (2004) and Bergek and Berggren (2014) mentioned in the introduction to this article; some of the stages identified are the same, the main difference being that our interview data show that these stages overlap and are not necessarily consecutive, and we highlight the main factors affecting each stage. This interpretation might be as well seen as aligned to the ideas of Jacobsson and Bergek (2011) on the systemic nature of the innovation process and its actors.

Previous studies highlighted the complexity of environmental innovation and the role of different factors. Hellsmark et al. (2016) perform 17 interviews and 2 stakeholder workshops covering Swedish cases of pilot and demonstration projects in the thermal conversion of black liquor and biochemical conversion of biomass and use "key analytical dimensions" that share similarities with our identified factors and stages, in particular the risk reduction and institutional precondition, and the role of actor networks. Söderholm et al. (2019) perform a similar analysis of Nordic technology development stages, focusing on the role of policies and networks in the process and arguing for the use of systemic instruments that support the functions operating at the innovation system level. Söderholm et al. (2019) stress the role of actors, networks, and institutions, elements that we have grouped under the *organization* factor, and the role of policies that we cover under the *market* factor. Ben Arfi et al. (2018) show the important role of knowledge sharing in green innovation, something that we have also found and discussed as part of both the *organization* and *technology* factors. Mazzucato and Semieniuk (2018) were able to quantify how financing affects the innovation process and its direction, for the case of renewable energies, a role that we can confirm for the technologies analyzed in this study and addressed under the *dealing with risk* stage.

Finally, one might discuss whether specific frameworks of innovation could have been used for the analysis of the interviews instead of deriving inductively our stage-based structure. For example, either the systems of innovations or transformative change frames described in detail by Schot and Steinmueller (2018) could have been guiding the analysis of the interview data. As specified in the method section, however, the rather classic stage-based framework allows to reconstruct the narrative behind each interview in the most accurate way as well as for a disaggregation of the data into meaningful units, and this was considered pragmatically the best choice. We cannot rule out that this might be a result of how the interview guide had been structured, focusing on the challenges experienced during technology development and inviting inventors to provide their experience and "story."

5 Conclusions

This study has investigated the fate of green technologies developed in the Nordic region from their conception to the market and identified the main factors that shape this process of environmental innovation, as well as specific stages during the process. The work proceeded using explorative qualitative research and was supported by the analysis of patents and interviews.

The work reconciled the experiences of more than 40 inventors in four Nordic countries under a coherent schematic narrative about environmental innovation in the Nordics. We are able to describe this process using a structure where four main factors (organization, technology, funding, and market) influence a technology life cycle of seven different stages (R & D, start-up, patenting, upscaling, dealing with risk, commercialization, and adoption).

A first insight based on this study confirms that the process of environmental innovation is indeed complex, and this complexity is worth embracing for a full understanding, especially when putting substantial expectations on technological solutions as part of sustainability transitions. However, we have reconstructed and analyzed the narratives and stories behind the development of various green technologies within a common structure, which is relatively simple and to some extent can even be considered rather linear but allows to analyze and understand in an organized way how different factors affect different stages of environmental innovation.

A second key insight from our analysis is that even in an environmentally ambitious policy context, as in the Nordics, upscaling of green technology innovations remains an almost impervious and difficult process, harder and more enduring than most inventors—and certainly the public as well as many policymakers—expect, and that there is a real "valley of death" for green technologies, which constitutes a major obstacle in terms of financing, and a "catch 22" problem for technology developers. These are significant major obstacles that need to be taken seriously and inform perceptions of how fast the transition to a low-carbon economy might be achieved. Given the urgency of the climate crisis and the need for deep emission reductions acknowledged with the Paris Agreement, our findings are not exactly encouraging with respect to the potentials of green growth for short-run achievements. Improved financing tools and targeted policies will be required to unleash the potentials of the technologies which are there in the labs and pilot experiments but too frequently stuck in the complexity of upscaling.

From here, further research could proceed in different directions. We noticed a certain segmentation in the type of inventors analyzed, and therefore, it might even be possible to further analyze environmental innovations from a gender perspective (Kaijser & Kronsell, 2014). Our interviews highlight that the inventors in environmental innovation are driven not only by commercial interests but also by ethical ones, but the same might not be the case for investors. Given the momentum currently existing in areas like sustainable finance and sustainable investments, the divestment movement, it is worth understanding further how ethical concerns (or the lack of) are incorporated in financing decisions regarding green technologies, from the investor side. Moreover, we find it important to extend the analysis to other green technologies beyond those included in this study and beyond those already extensively addressed in the literature (biofuels, photovoltaics, wind turbines) and to perform comparative analysis between technologies regarding the environmental innovation process and innovation in other areas, in particular information technologies, as this was often mentioned as a diametrically different example by our respondents. Our results indicate that there are more similarities than differences across green technologies and that results are reasonably generalizable across them. Given the challenges in making these technologies succeed at scale, we suggest concentrating public policies on identifying solutions for the funding challenges in the upscaling stage and to do further research investigating the interdependencies between green growth and green technology innovations.

The added value of this work is to provide a detailed and rich picture of the challenges and context in green technology developments within the Nordic countries, based on real-world cases of green technologies for the bio-based economy, for energy production and storage, and for reducing toxic emissions and nutrient loss. We believe that the insights from the concrete cases and the analysis will allow inventors, researchers, and policymakers in the domain of green and environmental technological innovation to better understand and critically evaluate the opportunities as well as challenges of technology-driven green growth, for improved technology development and policymaking, in the Nordic area and beyond.

Acknowledgments Financial support (grant #82841) from NordForsk, Nordic Energy, and Nordic Innovation to the Project NOWAGG (New Nordic Ways to Green Growth; https://projects.au.dk/nowagg/).

We extend our thanks to our NOWAGG project partners in particular Patrik Söderholm, Hans Hellsmark, and Roger Hildingsson for their constructive comments during the study design and implementation and to early versions of this manuscript.

We declare no competing interest.

References

Andersen, M. S. (2020). Small state syndrome or novel middle power? Climate leadership in governance, green finance and global engagement by the five Nordic council countries. In R. Wurzel (Ed.), *Multi-level and polycentric climate policy*. Routledge.

Beck, M., Lopes-Bento, C., & Schenker-Wicki, A. (2016). Radical or incremental: Where does R & D policy hit? Res. *Policy, 45*, 869–883. https://doi.org/10.1016/j.respol.2016.01.010

Ben Arfi, W., Hikkerova, L., & Sahut, J. M. (2018). External knowledge sources, green innovation and performance. *Technological Forecasting and Social Change*. https://doi.org/10.1016/j.techfore.2017.09.017

Bergek, A., & Berggren, C. (2014). The impact of environmental policy instruments on innovation: A review of energy and automotive industry studies. *Ecological Economics, 106*, 112–123. https://doi.org/10.1016/j.ecolecon.2014.07.016

Bowen, A., & Hepburn, C. (2014). Green growth: An assessment. *Oxford Review of Economic Policy, 30*, 407–422. https://doi.org/10.1093/oxrep/gru029

Braun, V., & Clarke, V. (2006). Using thematic analysis in psychology. *Qualitative Research in Psychology, 3*, 77–101. https://doi.org/10.1191/1478088706qp063oa

Breschi, S. (2000). The geography of innovation: A cross-sector analysis. *Regional Studies, 34*, 213–229. https://doi.org/10.1080/00343400050015069

Cantner, U., Graf, H., Herrmann, J., & Kalthaus, M. (2016). Inventor networks in renewable energies: The influence of the policy mix in Germany. *Research Policy*. https://doi.org/10.1016/j.respol.2016.03.005

Choi, H. (2018). Technology-push and demand-pull factors in emerging sectors: Evidence from the electric vehicle market. *Industry and Innovation*. https://doi.org/10.1080/13662716.2017.1346502

Costantini, V., Crespi, F., Martini, C., & Pennacchio, L. (2015). Demand-pull and technology-push public support for eco-innovation: The case of the biofuels sector. *Research Policy, 44*, 577–595. https://doi.org/10.1016/j.respol.2014.12.011

Cuerva, M. C., Triguero-Cano, Á., & Córcoles, D. (2014). Drivers of green and non-green innovation: Empirical evidence in low-tech SMEs. *Journal of Cleaner Production, 68*, 104–113. https://doi.org/10.1016/j.jclepro.2013.10.049
Dolfsma, W., & Leydesdorff, L. (2011). Innovation systems as patent networks: The Netherlands, India and nanotech. *Innovation: Management, Policy and Practice, 13*, 311–326. https://doi.org/10.5172/impp.2011.13.3.311
Dosi, G. (1982). Technological paradigms and technological trajectories: A suggested interpretation of the determinants and directions of technical change. *Research Policy, 11*, 147–162. https://doi.org/10.1016/0048-7333(82)90016-6
EPO. (2019). *Espacenet - Database of the European Patent Office* [WWW Document]. https://worldwide.espacenet.com/
Erlingsson, C., & Brysiewicz, P. (2017). A hands-on guide to doing content analysis. *African Journal of Emergency Medicine, 7*, 93–99. https://doi.org/10.1016/j.afjem.2017.08.001
Fevolden, A. M., & Klitkou, A. (2017). A fuel too far? Technology, innovation, and transition in failed biofuel development in Norway. *Energy Research and Social Science, 23*, 125–135. https://doi.org/10.1016/j.erss.2016.10.010
GGGI. (2014). *Green growth in practice: Lessons from country experiences.*
Gioia, D. A., Corley, K. G., & Hamilton, A. L. (2013). Seeking qualitative rigor in inductive research: Notes on the Gioia methodology. *Organizational Research Methods*. https://doi.org/10.1177/1094428112452151
Grubb, M. (2004). Technology innovation and climate change policy: An overview of issues and options. *Keio Economic Studies, 41*, 103–132.
Grunwald, A. (2018). Diverging pathways to overcoming the environmental crisis: A critique of eco-modernism from a technology assessment perspective. *Journal of Cleaner Production*. https://doi.org/10.1016/j.jclepro.2016.07.212
Haščič, I., & Migotto, M. (2015). *Measuring environmental innovation using patent data*. OECD Environment Working Papers, No. 89. Paris. https://doi.org/10.1787/5js009kf48xw-en
Hekkert, M. P., Suurs, R. A. A., Negro, S. O., Kuhlmann, S., & Smits, R. E. H. M. (2007). Functions of innovation systems: A new approach for analysing technological change. *Technological Forecasting and Social Change, 74*, 413–432. https://doi.org/10.1016/j.techfore.2006.03.002
Hellsmark, H., Frishammar, J., Söderholm, P., & Ylinenpää, H. (2016). The role of pilot and demonstration plants in technology development and innovation policy. *Research Policy, 45*, 1743–1761. https://doi.org/10.1016/j.respol.2016.05.005
Hickel, J., & Kallis, G. (2019). Is green growth possible? *New Political Economy, 1–18*. https://doi.org/10.1080/13563467.2019.1598964
Hildingsson, R., Kronsell, A., & Khan, J. (2019). The green state and industrial decarbonisation. *Environmental Politics*. https://doi.org/10.1080/09644016.2018.1488484.
Hojnik, J., & Ruzzier, M. (2016). What drives eco-innovation? A review of an emerging literature. *Environmental Innovation and Societal Transitions, 19*, 31–41. https://doi.org/10.1016/j.eist.2015.09.006
Investopedia. (2019). Green tech definition [WWW Document].
Jacobsson, S., & Bergek, A. (2011). Innovation system analyses and sustainability transitions: Contributions and suggestions for research. *Environmental Innovation and Societal Transitions, 1*, 41–57. https://doi.org/10.1016/j.eist.2011.04.006
Johnstone, N., Haščič, I., & Popp, D. (2010). Renewable energy policies and technological innovation: Evidence based on patent counts. *Environmental and Resource Economics, 45*, 133–155. https://doi.org/10.1007/s10640-009-9309-1
Kaijser, A., & Kronsell, A. (2014). *Climate change through the lens of intersectionality*. Polit. https://doi.org/10.1080/09644016.2013.835203
Karvonen, M., Kapoor, R., Ojanen, V., & Heinim, H. J. (2014). Commercialization of early stage university-based inventions. In *IEEE International Conference on Industrial Engineering and Engineering Management* (pp. 694–698). https://doi.org/10.1109/IEEM.2013.6962500

Kronsell, A. (2013). Gender and transition in climate governance. *Environmental Innovation and Societal Transitions, 7*, 1–15. https://doi.org/10.1016/j.eist.2012.12.003

Mazzucato, M. (2013). The entrepreneurial state: Debunking the public vs. private myth in risk and innovation. https://doi.org/10.1016/S0048-7333(01)00108-1

Mazzucato, M., & Semieniuk, G. (2018). Financing renewable energy: Who is financing what and why it matters. *Technological Forecasting and Social Change, 127*, 8–22. https://doi.org/10.1016/j.techfore.2017.05.021

Murphy, L. M., & Edwards, P. L. (2003). Bridging the valley of death: Transitioning from public to private sector financing. US Department of Energy/Nuclear Renewable Energy Laboratory.

Nemet, G. F. (2009). Demand-pull, technology-push, and government-led incentives for non-incremental technical change. *Research Policy, 38*, 700–709. https://doi.org/10.1016/j.respol.2009.01.004

Nemet, G. F., Zipperer, V., & Kraus, M. (2018). The valley of death, the technology pork barrel, and public support for large demonstration projects. *Energy Policy, 119*, 154–167. https://doi.org/10.1016/j.enpol.2018.04.008

Nielsen, H. Ø., & Pedersen, A. B. (2013). Hvordan kan staten fremme innovation, der fører til bæredygtige energisystemer? *Politica, 45*.

O'Neill, D. W. (2020). Beyond green growth. *Nature Sustainability, 3*, 260–261. https://doi.org/10.1038/s41893-020-0499-4

Palage, K., Lundmark, R., & Söderholm, P. (2019). The impact of pilot and demonstration plants on innovation: The case of advanced biofuel patenting in the European Union. *International Journal of Production Economics, 210*, 42–55. https://doi.org/10.1016/j.ijpe.2019.01.002

Peters, M., Schneider, M., Griesshaber, T., & Hoffmann, V. H. (2012). The impact of technology-push and demand-pull policies on technical change - Does the locus of policies matter? *Research Policy*. https://doi.org/10.1016/j.respol.2012.02.004

Scandura, A. (2019). The role of scientific and market knowledge in the inventive process: Evidence from a survey of industrial inventors. *The Journal of Technology Transfer, 44*, 1029–1069. https://doi.org/10.1007/s10961-017-9643-3

Schot, J., & Steinmueller, W. E. (2018). Three frames for innovation policy: R & D, systems of innovation and transformative change. *Research Policy, 47*, 1554–1567. https://doi.org/10.1016/j.respol.2018.08.011

Söderholm, P., Hellsmark, H., Frishammar, J., Hansson, J., Mossberg, J., & Sandström, A. (2019). Technological development for sustainability: The role of network management in the innovation policy mix. *Technological Forecasting and Social Change, 138*, 309–323. https://doi.org/10.1016/j.techfore.2018.10.010

Sovacool, B. K. (2017). Contestation, contingency, and justice in the Nordic low-carbon energy transition. *Energy Policy, 102*, 569–582. https://doi.org/10.1016/j.enpol.2016.12.045

The World Bank. (2012). Inclusive green growth: The pathway to sustainable development. *The World Bank*. https://doi.org/10.1205/psep.05009

Vaismoradi, M., Turunen, H., & Bondas, T. (2013). Content analysis and thematic analysis: Implications for conducting a qualitative descriptive study. *Nursing & Health Sciences, 15*, 398–405. https://doi.org/10.1111/nhs.12048

Ward, J. D., Sutton, P. C., Werner, A. D., Costanza, R., Mohr, S. H., & Simmons, C. T. (2016). Is decoupling GDP growth from environmental impact possible? *PLoS One*. https://doi.org/10.1371/journal.pone.0164733

Westholm, E., & Beland Lindahl, K. (2012). The Nordic welfare model providing energy transition? A political geography approach to the EU RES directive. *Energy Policy, 50*, 328–335. https://doi.org/10.1016/J.ENPOL.2012.07.027

Ziemnowicz, C. (2013). Joseph A. Schumpeter and innovation. In *Encyclopedia of creativity, invention, innovation and entrepreneurship*. https://doi.org/10.1007/978-1-4614-3858-8_476

Fiscal Incentives for Circular Economy: Insights from the Baltic States

Tarlan Ahmadov, Wolfgang Gerstlberger, and Gunnar Klaus Prause

Abstract The adoption of circular practices within environmental management is gaining worldwide recognition owing to the rapid resource depletion and the detrimental effects of climate change. Developments in Europe concerning the transition toward a circular economy (CE) have been significant, driven by both public and private initiatives, but with regional differences. To help the Baltic region catch up with Western Europe, this chapter provides a broad overview of economic instruments by answering the following question: "What are the existing fiscal incentives in the Baltic States for the enterprises in their transition towards Circular Economy?" To carry out this analysis, a desk research methodology was applied that involved collecting and analyzing the latest available data from each state's ministries and agencies dealing with taxation, subsidies, state aids, and government funds. At the same time, each state's trade policy was considered. Findings show the commitment of each state to pursue energy efficiency, research and development (R&D), information and communication technologies (ICT), and digitalization. The accumulation of knowledge allows policymakers, companies, and other stakeholders to better understand the dedication of the three states to a CE achievement process.

1 Introduction

Since the beginning of the industrial revolution, the economy has been shaped and dominated by a "take, make, and dispose" model. This way of production, known as a linear model, has finally reached the stage of no return, creating an imbalance between resource supply and goods demand (Murray et al., 2017). With the current trend in production and consumption rate, it will not be possible to sustain the development and ensure resources for the future generation. This is one of the most compelling issues of our time: a contradiction between the needs for economic

T. Ahmadov (✉) · W. Gerstlberger · G. K. Prause
Department of Business Administration, Tallinn University of Technology, Tallinn, Estonia
e-mail: tarlan.ahmadov@taltech.ee

V. Prokop et al. (eds.), *Business Models for the Circular Economy*, Sustainability and Innovation, https://doi.org/10.1007/978-3-031-08313-6_9

development and the necessities for accomplishing environmental sustainability (Hobson & Lynch, 2016; Marques et al., 2018).

The need for an alternative model of sustainable development with closed material loops, the circular economy (CE), emerged from the literature (Blomsma & Tennant, 2020; Salvatori et al., 2019; Tura et al., 2019). The definition of the CE concept has no clear boundary, but the idea behind it is to maintain the value of materials and keep them in circulation, thus reducing our reliance on material extraction (Kirchherr et al., 2017). According to Hislop and Hill (2011, p. 2): "The circular economy represents a development strategy that maximizes resource efficiency and minimizes waste production, within the context of sustainable economic and social development."

However, after decades of discussion about and research into CE, the linear model is still deeply entrenched. It is obvious that there are obstacles in political, economic, technological, and legal areas to the transition to CE (Hart et al., 2019). There is a strong need for political support to promote the concept of CE (Araujo Galvão et al., 2018). As supported by the literature (Prieto-Sandoval et al., 2018; Tura et al., 2019; de Jesus & Mendonça, 2018), a coordinated set of fiscal measures is needed to promote the inclusive transition to CE. Governments should play a leading role by reforming existing laws while passing new regulations and promoting the application of new environmental technologies and increasing public education (Kalmykova et al., 2018).

To the best of our knowledge, a study about the application of economic instruments at a macro level for an extended geographical area is still missing from the literature. For that reason, this study will concentrate on a regional area, specifically the Baltic States that are lagging behind their leading European counterparts, in order to support the transition and catch up on the policy framework. The accumulation of knowledge should allow policymakers, companies, and other stakeholders to better understand the dedication of the three states to a CE achievement process. The objective of this chapter is to provide insight into the Baltic States' fiscal incentives for CE by addressing the below question:

What are the existing fiscal incentives in the Baltic States that foster the business transition to CE?

To answer the research question, the study will gather gray literature by employing the desk research method. It will include data on state aid and state funds which aims to incentivize CE through state resources which may take on a variety of forms, including grants, interest, tax reliefs, and guarantees, and by providing goods and services on preferential terms (Aranda-Usón et al., 2019). Also, data from international organizations (e.g., OECD) as well as European schemes will be considered.

This chapter is organized as follows: review on the circular economy, followed by the definition and role of financial incentives to CE transition. The next section provides an overview of the Baltic States, followed by a description of the method. After the methodology, the results are summarized and discussed, and a conclusion is drawn in relation to policy suggestions, including a future research recommendation.

2 Conceptual Framework

2.1 Circular Economy

In 1990, two British environmental economists, D.W. Pearce and R.K. Turner, raised the concept of CE in their book *Economics of Natural Resources and the Environment* (Pearce & Turner, 1990). They described the traditional economy as having no in-built tendency to recycle, and the environment was perceived as a reservoir of wastes. In addition, the linear economy is considered to be a predominantly resource- and energy-intensive industrial model that responds to the capitalistic economic logic of unlimited growth of production and consumption (Smith, 2019). Yet, there was a need to develop a new economic model with a closed system where the environment and the economy are not characterized by linear interlinkages. The CE concept arose as a contrast to the linear economy, emphasizing the regenerative capacity of the ecosystem by specifically stressing the importance of minimizing the consumption of fossil fuel-based energy, prolonging the useful life of goods, and reusing all materials that enter the economic cycle in order to reduce waste and emissions (Geissdoerfer et al., 2017; Korhonen et al., 2018; Blomsma & Tennant, 2020).

At the EU level, the European Commission (2015, p. 2) has included a description of the CE concept in its Communication entitled "Closing the loop—An EU Action Plan for the circular economy," which is part of the Circular Economy Package. It provides an outline of policy and legislative measures for improved production, consumption, waste management, markets for secondary raw materials, innovation, investments, and monitoring. In this document, the emphasis on resource use is stressed, and CE is described as an economy "where the value of products, materials and resources is maintained in the economy for as long as possible, and the generation of waste minimised." The transition toward CE would make "an essential contribution to the EU's efforts to develop a sustainable, low carbon, resource-efficient and competitive economy." The economic dimension of CE is also prominent in this Communication which underlines that the transition toward CE is "the opportunity to transform our economy and generate new and sustainable competitive advantages for Europe." In this context, the EU Action Plan includes a series of measures aimed at addressing the full product cycle from production and consumption to waste management and the market for secondary raw materials.

From a systemic perspective, CE needs to decrease the actual ecological footprints to keep the economy within the physical limits or the carrying capacity of the planet (Rockström et al., 2009). To reduce the impact and keep the economic activities within the planetary limits, instruments for the promotion of this change of paradigm must involve diverse policies and actions at all levels. One of the essential instruments to enable and accelerate CE is a fiscal incentive (Di Maio et al., 2017; Elia et al., 2017). The next section clarifies what the fiscal incentives consist of and why they matter for the transition to CE. Understanding the view on fiscal incentives for CE helps to better grasp the idea and its potential benefits for the acceleration of the transition.

2.2 Fiscal Incentives as CE Transition Enabler

Fiscal incentives are features of fiscal policy, and they can influence and encourage the behavior of people and organizations to act in a particularly desirable way by offering a financial reward for certain activities. Financial incentives in the form of fiscal policies are designed to make certain decisions more appealing to people than they would otherwise be (Milios, 2021). Clear and consistent policies over a long period are needed if the governments are interested in encouraging the uptake of the CE concept among businesses. There needs to be a clear signal in the governments' policy, and it requires analysis of the potential opportunities and benefits that a CE approach could yield for businesses and economies. For example, if the government is interested in increasing the research and development (R&D) activities in a certain field, it could offer a tax credit to organizations that invest more money in R&D. If the interest of the government is to encourage the use of renewable energy in households, then the government could apply a variety of fiscal incentives, such as tax refunds on the investment in solar panels by households.

In connection with the transition to a circular economy, a large number of economic resources need to be invested in target sectors. A study by Studer et al. (2006) claims that lack of government incentives is one of the reasons for poor environmental management in an organization in Hong Kong. Tax or other financial incentives would drive organizations to engage more in environmental management. Findings from Tura et al. (2019) show high cost and lack of financial capability and support as barriers that prevent the transition toward CE. On the other hand, institutional support, such as incentives (e.g., for developing new solutions for cleaner production, for repairing, or for renovating), low-interest loans, subsidy policies, tax cuts, and refund policies, are driving factors for circular economy initiatives (Hart et al., 2019).

In the EU Member States (MS), environmental fiscal instruments and incentives, such as charges and taxes on waste, incineration, and landfills, tax reliefs and subsidies, or pay as you throw, are common measures. Government policies can affect the cost of production and the supply curve through taxes, regulations, and subsidies. In addition, there are other measures, such as tax rebates or premiums, that aim to incentivize the consumption of specific ecological products (Rabta, 2020).

From the firms' point of view, it is hard to introduce circular products as long as they need to compete with products derived from "tax-free" pollution because it is cheap to acquire and dispose of raw materials. Also, high labor costs are a burden in the transition for service-oriented business models as well as labor-intensive R&D efforts. Such barriers could be turned into a catalyst for the circular businesses if the taxation shifted from labor to consumption of natural resources and implemented "polluter pays" principles. The main goal here is to accelerate the growth based on human capital rather than the extraction of natural resources (Goovaerts & Verbeek, 2018).

To achieve the objectives of the promotion of CE, taxes must be imposed at the right level and must be directed at the source of the environmental burden. Several successful examples show that when taxes are implemented in this way, it is highly likely that the objectives will be achieved. Examples are the German energy and transport taxes; the Danish energy/carbon taxes; the UK climate change levy and fuel duty escalator; the Swedish NOx tax; the Finnish, Swedish, and UK waste taxes; the London congestion charge; and the Dutch wastewater effluent charge (Ekins et al., 2009).

Therefore, to reduce pollution and encourage the uptake of CE, the fiscal policy will be more efficient if such financial tools are applied. Considering the importance of institutional support, this study solely focuses on the existing fiscal incentives in the Baltic States for businesses to accelerate the transition toward a circular economy.

3 The Baltic States

The Baltic States, a group of three former Soviet Republics, Estonia, Latvia, and Lithuania, are located on the east coast of the Baltic Sea, in Northern Europe. Their total land area is about 175,000 km^2, and their populations count approximately 1.35, 1.9, and just under 2.8 million (Estonia, Latvia, and Lithuania, respectively). Since 2004, all three countries have become members of the European Union, their currency is the Euro, and they belong to NATO and OECD. They are listed in the Very High Human Development category by Human Development Index in 2020 (Estonia 0.892, Latvia 0.866, Lithuania 0.882) ("Human Development Index", 2020). The World Bank lists them among high-income level countries ("The World by Income and Region", 2020).

The Baltic States are seen as one group due to the similarity in terms of geography, size, economic structure, development, and demography, yet, with each their distinct culture. After regaining their independence from the Soviet Union in the early 1990s and joining the EU in 2004, their legal systems have benefited from the harmonization of legal norms. These countries form a closely integrated economic area.

While adapting to European standards, Estonia, Latvia, and Lithuania, each face its own set of challenges regarding the implementation and acceleration of the transition toward CE. This can be explained by the differences between the regions regarding their density, natural resources, industrial clusters, and economic or social factors. Since CE may look different in every region depending on local needs and resources, best practices from elsewhere cannot be effectively copy-pasted to the Baltic States. Every region in Europe should start with their own challenges to determine what the transition toward CE needs and build up experimental projects that can be scaled up and translated into policy. The industrial profile of a region plays an important role, with, for example, service- and resource-intensive sectors each calling for different types of support. The diversity of territorial contexts translates into different needs and opportunities that CE approaches should address.

Most importantly, after the adaptation of the Circular Economy Action Plan in 2015, the European Commission adopted a new package of deliverables in January 2018 (European Commission, 2018), including initiatives such as an EU strategy for plastics; a Communication on how to address the interplay between chemical, product, and waste legislations; a report on critical raw materials; and a framework to monitor progress toward a circular economy.

The development of circular economy strategies in EU member countries, together with EU frameworks, largely shapes the Baltic States' strategies on the implementation and transition to CE. In addition to the EU impact, any sort of investment to foster potential growth (e.g., transition to CE) in one of the three states in the Baltics would benefit the region as a whole as these countries grow together. Due to the close economic links between these countries, any development would spill over on their peers.

4 Methodology

To investigate the existing fiscal incentives for enterprises in the Baltic context, this chapter synthesized extant secondary data sources. More precisely, we reviewed the so-called gray literature provided by government entities from Estonia, Latvia, and Lithuania by employing desk research. It is seen as the appropriate research method for the collection of data from existing resources. It is often considered a low-cost method compared to field research. Table 1 shows a summary of the method used.

To start the data collection, the initial search was at EU level with inclusion (.eu sites) and exclusion (-news) criteria and with the below keywords:

"circular economy" AND "fiscal incentive" AND "European Union"

The initial search was done to understand the main frameworks that shape and guide the local policy at the state level as well as identify the key search words for a country-specific search. The below keywords were identified from the initial search to identify potentially relevant fiscal incentives at the state level:

Table 1 Summary of used research method

Research method	Desk research
Data collection method	Web-scan approach
Data analysis method	Descriptive analysis and qualitative content analysis
Sample	Websites of government-led or sponsored organizations
Period of data collection and analysis	June 2021–September 2021

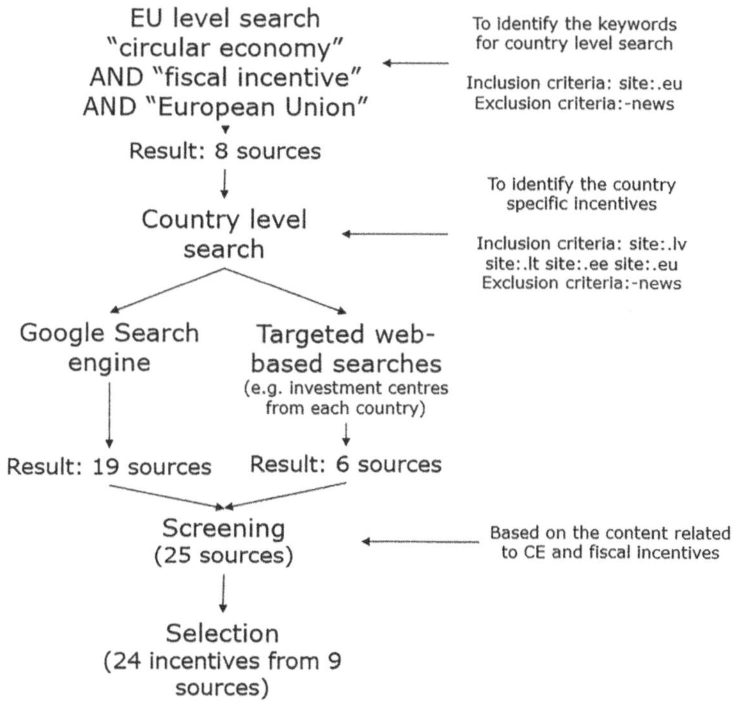

Fig. 1 The methodological procedure of the study (authors' own)

("Estonia" OR "Latvia" OR "Lithuania") AND ("sustainab*" OR "circul*") AND ("grant" / "tax" / "subsid*" / "support" / "incentive" / "fiscal" / "voucher" / "research" / "energy" / "material" / "innovation" / "green" / "efficien*")

The search technique has been refined stepwise (phrase searching, inclusion/ exclusion criteria) (Fig. 1). The collected data include fiscal instruments for supporting/promoting CE-related activities for businesses.

After the search, there was a screening where the executive summary of the relevant documents was carefully checked by the authors.

5 Findings

Table 2 provides an overview of the findings on selected 24 incentives from nine sources. The incentives were carefully analyzed in terms of content to elicit the fiscal incentives for CE as highlighted by the scientific literature. We organized incentives into three categories: resource efficiency, research and development (R&D), and information and communications technology (ICT) and digitalization. The sections below present and discuss the evidence on various fiscal incentives for CE transition from each Baltic State.

Table 2 Overview of the findings

Resource efficiency	Estonia	Adoption of new and innovative technologies https://kik.ee/et/toetatav-tegevus/innovaatiliste-tehnoloogiate-kasutuselevott-euroopa-liidu-innovatsioonifond (accessed on 18 November 2021)
		Development plans for heat management https://www.kik.ee/et/toetatav-tegevus/soojusmajanduse-arengukavad (accessed on 18 November 2021)
		Circular economy program https://kik.ee/et/toetatav-tegevus/ringmajanduse-programm (accessed on 18 November 2021)
		Resource efficiency for businesses https://www.kik.ee/et/toetatav-tegevus/ettevotete-ressursitohusus (accessed on 18 November 2021)
		The measure of the resource efficiency of enterprises https://ringmajandus.envir.ee/en/measure-resource-efficiency-enterprises (accessed on 18 November 2021)
	Latvia	Investments to establish or reconstruct the production premises to promote SMEs and energy savings reporting https://likumi.lv/ta/id/281590 (accessed on 12 November 2021)
		Investments to improve energy efficiency in food processing enterprises https://likumi.lv/doc.php?id=269868 (accessed on 12 November 2021)
		Efficient use of energy resources, reduction of energy consumption and transfer to renewable energy sources in the manufacturing industry https://likumi.lv/ta/id/310544 (accessed on 12 November 2021)
	Lithuania	Renewable energy resources for industry LT + https://www.esinvesticijos.lt/lt//finansavimas/kvietimai-teikti-paraiskas/paskelbti_kvietimai/kvietimas-teikti-paraiskas-finansuoti-projektus-pagal-priemone-atsinaujinantys-energijos-istekliai-pramonei-lt-nr-5-didelems-imonems (accessed on 12 November 2021)
		Eco consultant https://invega.lt/verslui/visos-priemones/25/eco-konsultantas-lt-98 (accessed on 11 November 2021)
		Eco-innovations LT + https://www.esinvesticijos.lt/lt//finansavimas/kvietimai-teikti-paraiskas/paskelbti_kvietimai/kvietimas-teikti-paraiskas-finansuoti-projektus-pagal-priemone-eco-inovacijos-lt-nr-4-1 (accessed on 11 November 2021)
R&D	Estonia	Green technology business development program https://www.kik.ee/en/supported-activity/green-technology-business-development-programme (accessed on 18 November 2021)
		Innovation voucher https://www.eas.ee/teenus/innovation-voucher/?lang=en (accessed on 18 November 2021)
		Development voucher https://www.eas.ee/teenus/development-voucher/?

(continued)

Table 2 (continued)

		lang=en#articleblock-Application (accessed on 18 November 2021)
	Latvia	Get-to-know-me voucher https://startuplatvia.eu/files/resources/editor/a4-innovation-vouchers-programme.pdf (accessed on 12 November 2021)
		Design voucher https://startuplatvia.eu/files/resources/editor/a4-innovation-vouchers-programme.pdf (accessed on 12 November 2021)
	Lithuania	Research and experimental development in the renewable energy sector https://www.e-mokslovartai.lt/funding-instruments/-/instrument/0465cab1-096f-429f-92e1-0702eaf92ffb# (accessed on 12 November 2021)
		InnoVouchers https://mita.lrv.lt/en/national-r-d-programmes/innovouchers (accessed on 11 November 2021)
		Green industry innovation https://www.e-mokslovartai.lt/all-calls-for-proposals/-/call/f89e1a92-63d5-4174-bf84-df7e8636c64c (accessed on 12 November 2021)
		EUROSTARS 2 https://www.e-mokslovartai.lt/all-calls-for-proposals/-/call/34a38c17-cf93-4fbe-a2dd-a7addaaca003 (accessed on 12 November 2021)
ICT and digitalization	Estonia	Digitalization grant https://www.eas.ee/teenus/digitalisation-grant/?lang=en (accessed on 18 November 2021)
		Green ICT https://www.eas.ee/teenus/norwaygrants/?lang=en (accessed on 18 November 2021)
	Latvia	Development of green industry innovation and ICT products https://www.liaa.gov.lv/en/about-us/norwegian-financial-mechanism-programme/open-call-for-small-grant-scheme (accessed on 12 November 2021)
		Application of green industry innovation and ICT products and technologies https://www.liaa.gov.lv/en/about-us/norwegian-financial-mechanism-programme/open-call (accessed on 12 November 2021)

5.1 Estonia

Among the three Baltic countries, Estonia is standing out due to its commitment to the number of fiscal incentives and the variety. The commitment from the state on resource efficiency, R&D, and ICT and digitalization is supported by the existing incentives to enterprises and is provided in the below tables.

Estonia has a number of incentives in the area of resource efficiency from various sources (Table 3) to encourage businesses to introduce resource-efficient measures by either implementation or development of new innovative solutions to increase resource productivity and efficient use of resources as well as improve resource efficiency by supporting activities, such as resource audit, implementation of resource-saving solutions, etc.

R&D activities in organizations related to CE in Estonia are supported by several incentives by the EU (Table 4). These incentives aim to catalyze the R&D activities in the development of new technologies that will contribute to a climate-neutral economy as well as improve the collaboration between industry and educational institutions, thus improving the competitive advantage of the companies.

Table 3 Fiscal incentives from Estonia to enterprises on resource efficiency

Adoption of new and innovative technologies	To increase the development and adoption of new and innovative low-carbon technologies with focus on production and use of renewable energy; the capturing, storage, and disposal of carbon; energy storage, including energy storage solutions in the industry	60% of the costs (max EUR 7.5 million)	European Union Innovation Fund
Development plans for heat management	To reduce the final consumption of energy based on more efficient production and transmission of heat energy	90% of the costs (max EUR 5000)	European Union Cohesion Fund
Circular economy program	To support activities that contribute to more efficient usage of the resources and help to introduce the principles of CE, prevent waste and emissions, and reduce the environmental impact of activities	30% construction projects and up to 50% for other projects	Minister of the Environment
Resource efficiency for businesses	To increase Estonia's resource productivity by developing and implementing resource-saving solutions, purchasing, and/or replacing equipment with innovative and more resource-efficient ones	50% of the costs (max EUR 200,000)	European Regional Development Fund
The measure of the resource efficiency of enterprises	To increase the resource productivity by the resource audit of an enterprise, which is a prerequisite for investment aid and investments in resource-efficient solutions	50% of the costs (max EUR 7500)	Ministry of the Environment, Structural Funds of the European Union

Table 4 Fiscal incentives from Estonia to enterprises concerning R&D

Green technology business development program	To develop activeness and entrepreneurship related to green tech, thus contributing to the move toward a climate-neutral economy and improving the competitiveness of Estonian green tech companies	90% of the costs (EUR 3000–40,000)	EU emissions trading system
Innovation voucher	To enable SMEs that are cooperating with a higher education institute, test laboratory, or intellectual property experts to develop innovative solutions for development obstacles, etc.	80% of the costs (max EUR 6000)	European Regional Development Fund
Development voucher	To support preliminary research that will enable the entrepreneur to gain comprehensive knowledge	70% of the costs (max EUR 35,000)	European Regional Development Fund

Table 5 Fiscal incentives from Estonia to enterprises concerning ICT and digitalization

Digitalization grant	To encourage enterprises in the industrial sector to invest in digital technologies, implementation of robots, and automation	10–50% of the cost (max EUR 200,000)	Ministry of Economic Affairs and Communications
Green ICT	To support value creation and sustainable growth in Estonia's private sector	50% of the costs (EUR 50,000–200,000)	Norway Grants

Estonia has a positive attitude toward cleantech and related innovation, and our search produced some incentives in this category (Table 5). As ICT is one of the enablers of eco-innovative solutions, incentives by Norway Grants to Estonian businesses aimed to increase value creation which could help Estonia to climb up in the global value chain. Another technology-related incentive, digitalization grants, could create an efficient and effective infrastructure and improve digitalization applications to improve the competitiveness of companies.

5.2 *Latvia*

In Latvia, the concept of CE is gaining recognition, but it is not the most debated topic among politicians or in society. The need to educate businesses on CE has received some attention, and an international conference has been held in Riga, the capital city of Latvia, in 2016. Also, the state is interested in financing some activities of the organizations to encourage the adoption of and transition to CE. Fiscal incentives from Latvia on resource efficiency, R&D, and ICT and digitalization are provided below.

Table 6 Fiscal incentives from Latvia to enterprises on resource efficiency

Investments to establish or reconstruct the production premises to promote SMEs and energy savings reporting	Cofinancing of investments takes place to establish new or reconstruct existing production premises and infrastructure to encourage development, economic activity, growth of productivity of SMEs, and energy savings in the manufacturing industry and ICT sector	35% (for large companies) and 45% (SMEs) of the costs (max EUR 1,166,000)	European Regional Development Fund
Investments to improve energy efficiency in food processing enterprises	To increase the efficiency of processing of agricultural products by promoting the development of competitive cooperation and sustainable agricultural production	35% (for large companies) and 40% (SMEs) of the costs (EUR 150,000–700,000)	European Agricultural Fund for Rural Development (EAFRD)
Efficient use of energy resources, reduction of energy consumption, and transfer to renewable energy sources in the manufacturing industry	The aim of the support is to promote energy efficiency, reduce energy consumption, and help companies in their transition to renewable energy sources in the manufacturing sector	45% (large companies), 55% (medium-sized enterprises), and 65% (small enterprises) of the costs (EUR 50,000–600,000)	European Union Structural Funds and the Cohesion Fund 2014–2020

In the domain of resource efficiency, the Ministry of Education and Science of Latvia has defined energy efficiency as a priority science sector alongside sustainable transport and knowledge-based bio-economy. Findings on this topic (Table 6) show the support from the state to its businesses for the establishment of new or reconstruction of existing infrastructure and promoting new solutions for energy efficiency. The ultimate goal of these incentives is to upgrade the existing production premises to increase efficiency, introduce innovation, and promote energy efficiency.

The Latvian state also dedicates some efforts to R&D activities in order to achieve CE. Findings from this country in Table 7 show the supports offered by the Investment and Development Agency of Latvia (LIAA) aimed at fostering collaboration between the industry and the research community, promoting R&D activities as well as R&D specifically for a new product, process, or strategy design for implementation of innovations in the organizations.

Norwegian Financial Mechanism provides financial support to organizations in Latvia to ease the purchase of new equipment, software, licenses, and patents in the

Table 7 Fiscal incentives from Latvia to enterprises concerning R&D

Get-to-know-me voucher	The incentive is to support activities such as research, prototyping, industrial design, and feasibility studies	EUR 5000	Investment and Development Agency of Latvia (LIAA)
Design voucher	Support shall cover design work for the development of a new product, process, or strategy for the implementation of innovations in the company	Up to 85% (EUR 5000)	Investment and Development Agency of Latvia (LIAA)

Table 8 Fiscal incentives from Latvia to enterprises concerning ICT and digitalization

Development of green industry innovation and ICT products	The objective is to increase the competitiveness for Latvian enterprises within the focus areas of green industry innovation and ICT by development of innovative green products or technologies as well as development of innovative ICT products or technologies	70% of the costs (EUR 10,000–130,000)	Norwegian Financial Mechanism
Application of green industry innovation and ICT products and technologies	The support is given to project promoters for the application of new products and technologies with reduced environmental impact into green industry innovation. Application of ICT products in the production process ensures digitalization or automation of the manufacturing processes in the area of ICT	45–55% of the costs (EUR 200,000–600,000)	Norwegian Financial Mechanism

area of green industry innovation with primary focus on improving the competitiveness of Latvian enterprises (Table 8). The aim is to help organizations with the introduction of new products, including new technology, and consequently reduce the environmental impact through the digitalization of production processes or automation of manufacturing processes.

5.3 Lithuania

The term CE is getting attention from the media and has been encountered in various media publications, thanks to the long-term work and impact of NGOs in Lithuania. Due to the lack of CE awareness in the business sphere, there are fewer fiscal incentives in Lithuania regarding CE than in Latvia and Estonia. The results of desk research from Lithuania show limited fiscal incentives, purely focusing on resource efficiency and R&D.

Table 9 Fiscal incentives from Lithuania to enterprises concerning resource efficiency

Renewable energy resources for industry LT +	Installation of energy production capacities using renewable energy sources (RES), development, and implementation of new technologies for more efficient use of RES in industrial enterprises	Up to EUR 500,000	EU Structural Funds
Eco consultant	To provide consultation to SMEs on more efficient use of resources, conservation of natural resources, implementation of eco-innovation, and similar issues	EUR 4000	EU Structural Funds
Eco-innovations LT +	Investments in tangible assets (equipment, technologies) are planned to reduce environmental impact and/or conserve natural resources, waste-free production, reuse and/or recycling of waste, recovery of heat, etc.	35% (medium-sized enterprises) and 45% (small enterprises) of the costs (EUR 50,000–1,500,000)	EU Structural Funds

The incentives in Table 9 show the Lithuanian state's commitment to resource efficiency by offering financial support for organizations to invest in tangible assets in order to reduce the environmental impact as well as efficient use of renewable energy sources. In addition to assets, EU Structural Funds also provide support to SMEs with consultation on the establishment of their resource efficiency strategies.

On R&D, the positive attitude by the Lithuanian government can be seen from the findings (Table 10). Existing fiscal incentives encourage enterprises to collaborate and establish initial contacts with academia to develop and commercialize new technological solutions with the help of digitalization or other new products, services, or processes, in order to improve the competitiveness of Lithuanian enterprises. Overall, the mentioned fiscal incentives are the primary support tools that are directly targeting businesses introducing innovative solutions to help the transition toward CE through R&D activities.

6 Discussion

Most studies in CE are case- or industry-specific and focus on business model elements or business practices (e.g., Blomsma & Tennant, 2020; Di Maio et al., 2017) and enablers and barriers to deepening literature on the implementation and acceleration of the transition toward CE (e.g., Tura et al., 2019; Hart et al., 2019; de Jesus & Mendonça, 2018; Araujo Galvão et al., 2018). Despite the extensive

Table 10 Fiscal incentives from Lithuania to enterprises concerning R&D

Research and experimental development in the renewable energy sector	The aim of the support is energy development, energy efficiency, research, and experimental development in the renewable energy sector	Max of EUR 500,000	Agency for Science, Innovation and Technology (MITA)
InnoVouchers	To encourage businesses and science to establish initial contacts, to continue research and (or) experimental development activities	Up to EUR 4,500,000	European Regional Development Fund
Green industry innovation	Development and commercialization of new green products/technologies (e.g., with the help of digitalization, automation, robotization)	EUR 10,000–200,000	Norwegian Financial Mechanism
EUROSTARS 2	To support SMEs that carry out research and invest in innovation and development of new products, services, and processes to gain a competitive advantage	75% (medium-sized enterprises) and 80% (small enterprises) of the costs (max EUR 300,000)	Agency for Science, Innovation and Technology (MITA)

literature on CE, the literature has made no substantial progress in institutional support, specifically on fiscal incentives.

However, governments and businesses have grown increasingly aware of the CE concept, and our findings show some CE practices are already used as incentives by each state in certain contexts, through various instruments (such as resource efficiency and R&D). At the same time, it is clear that incentives for CE practices are not universal and each country is distinct in terms of opportunities and challenges (natural resources, industrial clusters, and social factors) in the transition toward CE (Salvatori et al., 2019). Investigating the fiscal incentives in each state helps us to understand the differences between the three countries in their efforts toward acceleration of CE and to propose measures enhancing institutional support, by removing barriers, and sharpening or enhancing drivers. The aim of this discussion is to provide stakeholders in each state with an overview of the current situation and possible solutions to transform the regional economies to more circular models.

While reviewing the findings, one can see that each state puts efforts into resource efficiency which are intended to contribute to the development of renewable energy resources as well as efficient use of resources. This may result from the fact that energy efficiency was one of the main environmental problems reported by the EU. In addition to EU, internationally, energy-related activities are the primary focus for policies and support programs (Durst & Gerstlberger, 2020). In the case of ICT and digitalization, several initiatives show the dedication of Estonia in this category, while Lithuania only focuses on the R&D activities that help to implement digitalization/automation, hereunder some energy efficiency aspects. Lithuania increasingly recognizes the need to provide consultation and advisory support to operators to increase voluntary compliance; however, guidance on good environmental practices is largely lacking.

The Baltic region also collaborates with other regions in Europe, especially with the Nordic countries. Currently, ongoing projects, such as BioBaltic, are one example of Nordic-Baltic cooperation regarding CE. The project aims to generate awareness of different bio-economy models through peer-to-peer learning and the building of networks across the Baltic and Nordic countries. This collaboration will enable knowledge generation and exchange on different aspects of bio-economy transition, including financing aspects, industrial partnerships, and symbiosis or opportunities of digitalization ("BioBaltic", 2021).

The number of EU Ecolabel products and Eco-Management and Audit Scheme (EMAS) licensed organizations in a country can provide a rough understanding of the measurements of the CE transition. As of September 2021, Estonia had 974 EU Ecolabel products, while Latvia and Lithuania had 79 and 474, respectively, out of 83,590 in the EU as such ("Ecolabel facts and figures", 2021). This figure shows the significant take-up of the labels by Estonia and indicates the extent to which the private sector is engaged in the transition.

If we look at Estonia separately, it is worth mentioning that the country is one of the pioneers in establishing an institute of engineering and circular economy (TTK) that is teaching specific skills in order to accelerate the transition toward CE. There are several other measures, such as the establishment of Eesti Pandipakend in 2005, a producer responsibility organization for recycling of packaging marketed by producers, importers, and traders in Estonia. Moreover, the Environmental Investment Centre (Keskkonnainvesteeringute Keskuse—KIK) was established in Estonia to finance environmental activities and investment projects. It is an agency handling implementation of aid from the state and EU funds. Clear guidance and comprehensive information are available on their website for businesses that are seeking various types of financial support. In addition to their website, information on incentives is published in national newsletters and on the website of the Ministry of Environment.

In addition to the Investment Centre, Estonia has created opportunities such as the Circular Economy Forum and the Estonian Association for Environmental Management for the development of networks among all stakeholders involved in CE, in order to increase awareness, share information, and promote dialogues aimed at creating a national CE strategy. Also, the Estonian government is currently formulating a strategy and action plan for CE (Ringmajandus, 2021). However, in principle, the Ministry of Rural Affairs, the Ministry of the Environment, and the Ministry of Economic Affairs and Communication are all responsible for setting and implementing CE policies. This could create some challenges for the implementation of comprehensive circular policies.

The findings of the desk research show the numerous fiscal incentives related to resource efficiency for enterprises; however, Estonia is not ranked as the leading in the EU in the domain of resource efficiency as the country falls below the EU average due to its domestic material productivity (Ministry of Finance, 2018). The country has decided to invest EUR 111 million in more resource-efficient solutions, mainly in small- and medium-sized enterprises (SMEs) in the manufacturing industry (European Commission, 2019) which might catalyze the transition in this direction.

On the topic of R&D, in addition to incentives, Estonia signed a Memorandum of Understanding regarding an Energy Research Program between the Baltic States and the Nordic Energy Research on Energy Program in 2018 (IEA, 2019) to foster energy technology research and development. This memorandum could help enterprises to access more R&D activities in the sphere of energy efficiency. However, findings from OECD (2019) show that ICT-related patents make up a rather small share of all national patents filed. Also, automation in SMEs in more traditional sectors has been only poorly taken up.

A recent study by the Tallinn University of Technology (TalTech, 2022) shows that there is some level of awareness about the CE application in organizations and that there is a need for more financial and nonfinancial (consultancy, dissemination of best practices) support from the authorities. Financial support is needed to help organizations implement new technologies and build a cooperative network for the application of CE practices. Even though there is some awareness, the notion of CE is still very new in Estonia. Policy initiatives have only just started to prioritize this area specifically. The lack of a comprehensive policy framework for CE is a barrier to achieving sustainable use of resources throughout the entire product value chain. An overall policy strategy for material productivity and resource efficiency would strengthen the coordination of efforts across different sectors.

Desk research results from Latvia highlight the commitment of the state to CE-related activities. They show that the state has an interest in accelerating entrepreneurial activities related to resource efficiency, R&D, as well as ICT and digitalization. Latvia has a policy strategy named Guidelines for Research, Technology Development and Innovation 2014–2020 which outlines targets and strategies for research development and innovation. It also shows the state's support to businesses in our findings concerning fiscal incentives.

In October 2019, the Ministry of Environmental Protection and Regional Development (VARAM) published a draft informative report on the CE strategy of the state which is one of the first steps toward CE. Furthermore, transition to CE in the case of Latvia is likely to accelerate due to the inclusion of low carbon, climate-resilient development and sustainable use of natural resources in the National Development Plan for 2021–2027 (CSCC, 2022). It is the backbone of further policymaking and investment planning.

In recent years, academic interest from Lithuania regarding the topic of CE has been increasing. It is partly due to the increase in awareness-raising efforts by NGOs and other organizations. One of the most notable examples of an NGO is "Žiedinė ekonomika" which has been actively covering new developments within CE, presenting recommendations to the Lithuanian government. Similar NGOs cover the topics of how circular practices can be adopted, the challenges that Lithuania faces, and what can be done to eliminate the barriers. In addition to NGOs, informal education opportunities, such as the "Man rūpi rytojus" project, offer circular economy principles to citizens, policymakers, etc. To increase public awareness on CE, the municipality of the second largest city in Lithuania, Kaunas, purposefully designed the annual Christmas tree from recyclable and recycled materials. As the country is quite committed in terms of public awareness regarding the circular economy, this is also mirrored in the business sector. The "EKODA" association for ecological design promotes sustainable design principles to Lithuanian designer works.

There are two notable strategic policy frameworks in Lithuania in the energy sector which are partly relevant to CE. The Law on Energy from Renewable Sources aims to ensure sustainable development of renewable energy sources, and the National Programme for the Development of Renewable Energy Sources 2017–2023 sets national targets for and promote the use of local renewable energy sources. These policy frameworks also turn up in the findings on the fiscal incentive for businesses.

The country might still lack the specific policy frameworks that support the transition toward CE, however, work done by NGOs and various media channels influence the existing policies. As it could be seen from the findings on Lithuania, incentives are concentrated on resource efficiency and R&D. Lithuania is still missing a document that specifically targets CE policy for the country. Overall, little government involvement continues to hinder CE attempts in Lithuania.

Overall, to steer the transition toward CE and guide related investment choices, Latvia and Lithuania need to further broaden the cooperation across ministries and with stakeholders and consider establishing a dedicated institutional platform.

7 Conclusion

This study has reviewed the existing fiscal incentives for CE in the Baltic States using the desk research method. Overall, the Baltic countries form a closely integrated economic area, and the economy keeps growing since the restoration of independence from Soviet occupation. All three countries continue their economic development in the traditional industries rather than the circular economy. Each state's policy framework is relying heavily on EU frameworks. As the trend in the EU is in favor of CE implementation in the public and private sector, the three Baltic countries are also encouraging the transition to CE through different policy frameworks as well as fiscal incentives.

While the European Commission is creating guidelines and frameworks to accelerate the transition to CE, some countries have already taken their own set of actions in order to create favorable conditions for circular business models. At the same time, it is important not to underestimate territorial differences in terms of opportunities and challenges in the transition toward CE. Natural resources, industrial clusters, as well social factors play a key role, which is also visible from the findings. Among the Baltic countries, Estonia appears to be most likely to become sustainable in its economic activities due to its geographical and cultural similarity to Finland. All three countries' commitment is reflected in the national economic strategies, and all three countries have been supported by Norway Grants which have an important impact on green technology development.

Having gained insight into each country's commitments to CE, we were able to draw up some recommendations. Some examples are as follows: To foster the transition towards a circular economy, policy framework needs to be developed and strengthened. There must be diverse support mechanisms in addition to institutional support. Implementation support programs need to be developed for companies in order to help them engage more in EU funding. The practical application of state-level incentives for enterprises must be improved. The public engagement in company compliance on circularity must be strengthened. Structured and comprehensive reports on industry achievements for more company-level engagement must be published.

The unavoidable methodological and database limitations of this chapter point to avenues for further research. First, whereas this chapter adopted a desk-based research approach, other techniques such as interviews with state representatives from relevant ministries or investment centers may be able to take this study further through collection of firsthand data and insights. Second, the needs of entrepreneurs in terms of institutional support must be identified through an in-depth study. Such a study will help policymakers to target the needs and gaps of organizations by means of direct fiscal and other incentives. Third, the chapter has a limitation of focus as it is limited to the Baltic States. To overcome this limitation, a cross-regional study might be conducted to generalize critical insights into the role of fiscal incentives for CE. Finally, a deeper understanding is needed regarding the impact of fiscal incentives on the organizations in relation to implementation of CE practices. This could be studied longitudinally by focusing on the organizations that have received fiscal incentives for the planned activities regarding CE. In this sense, the insights that are provided in this chapter facilitate the development of policy guidelines.

References

Aranda-Usón, A., Portillo-Tarragona, P., Marín-Vinuesa, L., & Scarpellini, S. (2019). Financial resources for the circular economy: A perspective from businesses. *Sustainability, 11*(3), 888. https://doi.org/10.3390/su11030888

Araujo Galvão, G., de Nadae, J., Clemente, D., Chinen, G., & de Carvalho, M. (2018). Circular economy: Overview of barriers. *Procedia CIRP, 73*, 79–85. https://doi.org/10.1016/j.procir.2018.04.011

BioBaltic. Nordregio projects. (2021). Retrieved January 8, 2022, from https://nordregioprojects.org/biobaltic/

Blomsma, F., & Tennant, M. (2020). Circular economy: Preserving materials or products? Introducing the resource states framework. *Resources, Conservation & Recycling, 156*, 104698. https://doi.org/10.1016/j.resconrec.2020.104698

Creating a strategy and action plan for circular economy in Estonia (Ringmajandus). (2021). Retrieved February 14, 2022, from https://ringmajandus.envir.ee/en/creating-strategy-and-action-plan-circular-economy-estonia.
Cross-Sectoral Coordination Center (CSCC). National Development Plan of Latvia for 2021–2027. (2022). Retrieved November 18, 2021, from https://www.pkc.gov.lv/sites/default/files/inline-files/NAP2027__ENG.pdf
de Jesus, A., & Mendonça, S. (2018). Lost in transition? Drivers and barriers in the eco-innovation road to the circular economy. *Ecological Economics, 145*, 75–89. https://doi.org/10.1016/j.ecolecon.2017.08.001
Di Maio, F., Rem, P., Baldé, K., & Polder, M. (2017). Measuring resource efficiency and circular economy: A market value approach. *Resources, Conservation & Recycling, 122*, 163–171. https://doi.org/10.1016/j.resconrec.2017.02.009
Durst, S., & Gerstlberger, W. (2020). Financing responsible small- and medium-sized enterprises: An international overview of policies and support programmes. *Journal of Risk and Financial Management, 14*(1), 10. https://doi.org/10.3390/jrfm14010010
Ecolabel facts and figures. European Commission. (2021). Retrieved November 14, 2021, from https://ec.europa.eu/environment/ecolabel-facts-and-figures_en
Ekins, P., Dresner, S., Potter, S., Shaw, B., & Speck, S. (2009). *The case for green fiscal reform: Final report of the UK Green Fiscal Commission*. Green Fiscal Commission.
Elia, V., Gnoni, M., & Tornese, F. (2017). Measuring circular economy strategies through index methods: A critical analysis. *Journal of Cleaner Production, 142*, 2741–2751. https://doi.org/10.1016/j.jclepro.2016.10.196
Enablers and barriers to the introduction of circular economy practices. TalTech. (2022). Retrieved February 6, 2022, from https://taltech.ee/en/news/study-circular-economy-application-completed-taltech-department-business-administration
European Commission. (2015). *Closing the loop – An EU action plan for the circular economy*. European Commission.
European Commission. (2018). *Communication on a monitoring framework for the circular economy*. European Commission.
European Commission. (2019). *The EU Environmental Implementation Review 2019. Country report – Estonia*. European Commission.
Geissdoerfer, M., Savaget, P., Bocken, N., & Hultink, E. (2017). The circular economy – A new sustainability paradigm? *Journal of Cleaner Production, 143*, 757–768. https://doi.org/10.1016/j.jclepro.2016.12.048
Goovaerts, L., & Verbeek, A. (2018). Sustainable banking: Finance in the circular economy. In *Investing in resource efficiency* (pp. 191–209). Springer.
Hart, J., Adams, K., Giesekam, J., Tingley, D. D., & Pomponi, F. (2019). Barriers and drivers in a circular economy: The case of the built environment. *Procedia CIRP, 80*, 619–624. https://doi.org/10.1016/j.procir.2018.12.015
Hislop, H., & Hill, J. (2011). *Reinventing the wheel: A circular economy for resource security*. Green Alliance.
Hobson, K., & Lynch, N. (2016). Diversifying and de-growing the circular economy: Radical social transformation in a resource-scarce world. *Futures, 82*, 15–25. https://doi.org/10.1016/j.futures.2016.05.012
Human Development Index. United Nations Development Programme. (2020). Retrieved November 10, 2021, from https://hdr.undp.org/en/2020-report
International Energy Agency. (2019). Energy policies of IEA countries: Estonia 2019 review, Brussels: IEA. Available online: https://www.iea.org/reports/energy-policies-of-iea-countries-estonia-2019-review (accessed on 10 November 2021).
Kalmykova, Y., Sadagopan, M., & Rosado, L. (2018). Circular economy – From review of theories and practices to development of implementation tools. *Resources, Conservation & Recycling, 135*, 190–201. https://doi.org/10.1016/j.resconrec.2017.10.034

Kirchherr, J., Reike, D., & Hekkert, M. (2017). Conceptualizing the circular economy: An analysis of 114 definitions. *Resources, Conservation & Recycling, 127*, 221–232. https://doi.org/10.1016/j.resconrec.2017.09.005

Korhonen, J., Honkasalo, A., & Seppälä, J. (2018). Circular economy: The concept and its limitations. *Ecological Economics, 143*, 37–46. https://doi.org/10.1016/j.ecolecon.2017.06.041

Marques, A., Fuinhas, J., & Pais, D. (2018). Economic growth, sustainable development and food consumption: Evidence across different income groups of countries. *Journal of Cleaner Production, 196*, 245–258. https://doi.org/10.1016/j.jclepro.2018.06.011

Milios, L. (2021). Towards a circular economy taxation framework: Expectations and challenges of implementation. *Circular Economy and Sustainability, 1*(2), 477–498. https://doi.org/10.1007/s43615-020-00002-z

Ministry of Environmental Protection and Regional Development (VARAM). Informatīvais ziņojums "Aprites ekonomikas stratēģija Latvijai". (2019). Retrieved November 14, 2021, from https://www.varam.gov.lv/sites/varam/files/content/files/varamzin_-200120_aestrategija.pdf

Ministry of Finance. (2018). FAKTILEHT - Riigikantselei ja Rahandusministeeriumi taustamaterjal EESTI 2035 valdkondlikele töörühmadele november-detsember 2018. Ministry of Finance.

Murray, A., Skene, K., & Haynes, K. (2017). The circular economy: An interdisciplinary exploration of the concept and application in a global context. *Journal of Business Ethics, 140*(3), 369–380. https://doi.org/10.1007/s10551-015-2693-2

OECD. (2019). *OECD economic surveys: Estonia*. OECD Publishing. Available online: https://www.oecd.org/economy/surveys/estonia-2019-OECD-economic-survey-overview.pdf (accessed on 10 November 2021).

Pearce, D. W., & Turner, R. K. (1990). *Economics of natural resources and the environment*. Harvester Wheatsheaf.

Prieto-Sandoval, V., Jaca, C., & Ormazabal, M. (2018). Towards a consensus on the circular economy. *Journal of Cleaner Production, 179*, 605–615. https://doi.org/10.1016/j.jclepro.2017.12.224

Rabta, B. (2020). An economic order quantity inventory model for a product with a circular economy indicator. *Computers & Industrial Engineering, 140*, 106215. https://doi.org/10.1016/j.cie.2019.106215

Rockström, J., Steffen, W., Noone, K., Persson, Å., Chapin, F., Lambin, E., et al. (2009). Planetary boundaries: Exploring the safe operating space for humanity. *Ecology and Society, 14*(2). https://doi.org/10.5751/es-03180-140232

Salvatori, G., Holstein, F., & Böhme, K. (2019). *Circular economy strategies and roadmaps in Europe: Identifying synergies and the potential for cooperation and Alliance building—Study* (1st ed., p. 2019). Brussels, Belgium.

Smith, R. (2019). An ecosocialist path to limiting global temperature rise to 1.5 C. In J. Morgan (Ed.), *Economics and the ecosystem*. World Economic Association Books.

Studer, S., Welford, R., & Hills, P. (2006). Engaging Hong Kong businesses in environmental change: Drivers and barriers. *Business Strategy and the Environment, 15*(6), 416–431. https://doi.org/10.1002/bse.516

The World by Income and Region. The World Bank. (2020). Retrieved November 9, 2021, from https://datatopics.worldbank.org/world-development-indicators/the-world-by-income-and-region.html

Tura, N., Hanski, J., Ahola, T., Ståhle, M., Piiparinen, S., & Valkokari, P. (2019). Unlocking circular business: A framework of barriers and drivers. *Journal of Cleaner Production, 212*, 90–98. https://doi.org/10.1016/j.jclepro.2018.11.202

Do Environmental Regulations Stimulate Firms' R&D, Product Innovation, or Environmental Awareness? Putting Porter's Hypothesis in the Context of Central and Eastern European Countries

Viktor Prokop, Jan Stejskal, and Cali Nuur

Abstract This chapter is based on analyzing Porter's hypothesis stating that pollution could be seen as a kind of resource waste and that a firm's reuse of this waste could lead to the firm's competitive advantage, higher R&D, product innovation, and performance. However, countries in the Europe differ due to their historical legacy. Using data from the World Bank's Enterprise Survey (2019), we put Porter's hypothesis in the context of the "catching-up" countries of Central and Eastern Europe (the Czech Republic, Slovakia, Poland, Estonia, Latvia, and Lithuania). These countries differ significantly from their Western neighbors, especially in terms of lower innovation performance and environmental awareness. The results of our analyses confirm the importance of environmental regulations and their influence on firms' R&D but also show them to be triggers for environmental awareness among firms. By contrast, we do not confirm the validity of the strong version of Porter's hypothesis, which expects environmental regulations to impact the performance of firms in the countries studied. In addition, we point out an important role of firms' R&D within product innovation processes. This chapter brings several contributions. From the theoretical point of view, we link theory regarding the validity of Porter's hypothesis on the "catching-up literature" and on firms' environmental awareness. From the practical perspective, we provide practical implications.

V. Prokop
Science and Research Centre, University of Pardubice, Pardubice, Czech Republic
e-mail: viktor.prokop@upce.cz

J. Stejskal (✉)
Institute of Economic Sciences, University of Pardubice, Pardubice, Czech Republic
e-mail: jan.stejskal@upce.cz

C. Nuur
Department of Industrial Economics and Management (INDEK), The Royal Institute of Technology, Stockholm, Sweden

V. Prokop et al. (eds.), *Business Models for the Circular Economy*, Sustainability and Innovation, https://doi.org/10.1007/978-3-031-08313-6_10

1 Introduction

The growing competition in world markets is putting pressure on individual market players to be constantly innovative and creative. The same is true for the growing population, which leads to increased consumption and a demand for new products. This inevitably leads to the depletion of nonrenewable resources, as well as to air and nature pollution. It provokes a reaction in society, which is aware of the need for action leading to environmental protection. Many economic actors therefore seek to abandon traditional, linear extract-produce-use-dump material and energy flow models (Korhonen et al., 2018), which were characteristics of existing ("modern") economic systems, in favor of models based on the (looped) circular economy principle, where eco-innovation and more sustainable products are created. This should result in the emergence of such a system, including alternative models that are cyclical and have regenerative designs (Geissdoerfer et al., 2017), underpinned by institutional capital (see, for example, Chizaryfard et al., 2021) that facilitate the emergence of a circular economy. To support these initiatives, countries have developed environmental regulations to control environmental challenges. Environmental regulations can be described as the set of standards that can directly and/or indirectly be implemented by institutions or governments to control environmental problems (Javeed et al., 2021).

The role of the public sector and its (environmental) regulations in this case has been a matter of recent years. This is due to differing views on the effects of these regulations. From one perspective, environmental regulations can restrict the production potential of firms and industries because of increased expenditures, such as taxes, levies, monitoring or administrative cost, and increased investment in green technologies (Böhringer et al., 2012). By contrast, however, properly designed and well-established environmental regulations can have great stimulating power that economic entities can induce behavioral change and enhance their competitiveness (Ambec & Barla, 2002; Huiban et al., 2018). This paradigm is known as the Porter hypothesis (Porter, 1991; Porter & Van der Linde, 1995). The essence of Porter's hypothesis is that firms may be beneficiaries of environmental regulations and that environmental regulation can trigger innovation and thus offset some or all compliance costs; this is referred to as the "innovation offset effect" (Bu et al., 2020). However, interestingly, the authors have not yet agreed on whether the Porter hypothesis can be generally recognized as valid, as the empirical evidence for these arguments is somewhat mixed (as shown in Sect. 2.1). Therefore, the validity of the Porter hypothesis is still one of the most controversial issues in the debate on environmental regulation, thus creating a research gap calling for additional studies in this area (Ambec et al., 2013; Huiban et al., 2018).

The first purpose of this chapter is therefore to contribute to the ongoing discussion about the validity of the Porter hypothesis for the example of selected countries from Central and Eastern Europe (CEE), which significantly differ from Western and Northern European countries (Prokop et al., 2021a). These differences can be observed both at the innovative level and at the environmental level. From the perspective of innovation performance, these are countries that achieve mostly average, but also below-average, innovation performance, for example, in the European Commission's European Innovation Performance Scoreboard measurements (European Commission, 2021). Reasons for the weaker innovation performance of these countries include, for example, insufficiently built infrastructure; mental lock-in; lower trust and conflicts of interest among cooperating partners, especially between universities and firms; and the inability to profit from research and development and from the provision of national and European public funds (Kotkova Striteska & Prokop, 2020). For these and many other reasons, these countries are considered catching-up states which, despite efforts to apply successfully functioning Western and Northern innovation models, still face higher levels of innovation inefficiency.

From an environmental perspective, the difference between CEE countries and Western and Northern European countries is mainly due to the historical context of CEE countries. The socialist regimes did not consider environmental issues fundamental in the past and did not solve them systematically. This has led both to environmental pollution and to a lower perception of environmental problems in society, which has emphasized the need for a quality, effective, and targeted regulatory function of the state (Horbach, 2016). Ever since the first attempts to solve these problems were observed in the 1990s, the role of the state and its regulations in environmental protection in CEE countries has remained a topical issue. Referring to Biscione et al. (2021), only a few previous contributions to the literature investigated the determinants of eco-innovation in these countries (as compared to high-income Western European countries). Moreover, to the best of our knowledge, current research still lacks a more comprehensive study of the validity of Porter's hypothesis in Central and Eastern Europe.

Following the above arguments that countries of Central and Eastern Europe are associated with lower environmental awareness, on both the supply (firms) and the demand (population) side, the second purpose of this chapter is to test how environmental regulations affect firms' environmental awareness in CEE countries. Moreover, referring to Kim (2015), who states that prior research primarily described the relation between firms' "voluntary activities," such as environmental awareness, and firms' sustainable performance but paid less attention to the relationship between these activities and the "non-environmental" performance of firms, we also test whether environmental awareness in CEE countries affects firms' R&D, innovation, and possibly (indirectly) also their performance.

This chapter contributes to both theory and practice in several ways. From the theoretical perspective, we contribute to the literature on the validity of the Porter hypothesis and create a comprehensive overview of studies examining the issue (see Sect. 2.1). The next theoretical contribution of this chapter is to the theory of catching-up (transition) CEE countries, which have so far received less attention in the literature than, for example, the countries of Western Europe. From a practical perspective, the main contribution of this chapter is to define the specific implications for firms and public policy-makers, especially in CEE countries. These are, for example, focused on increasing environmental awareness both in firms and in society and on the role of policy-makers within this process. As we affirm the key role of firms' R&D, our implications also focus on supporting their research capabilities and their absorption capacity. In addition, because our analysis works with an aggregated data set that includes 2139 firms from 6 CEE countries (Czech Republic, Slovakia, Poland, Estonia, Latvia, and Lithuania), another contribution of this study may be a greater possibility of generalizing our results (e.g., compared to country-specific studies).

The remainder of this chapter is as follows: Section 2 is divided into three parts and includes an overview of differing views on the Porter hypothesis and its validity; a theoretical overview of the link between environmental awareness, regulations, and firm performance; and the definition of the research question for CEE countries. Section 3 describes our research data, model, and methods. Results are shown in Sect. 4, and a discussion with implications is in Sect. 5. Finally, in Sect. 6, we provide conclusions, including proposals for future research directions.

2 Theoretical Background

2.1 An Overview of Differing Views on the Porters' Hypothesis and Its Validity

Conventional wisdom based on standard economic models suggests that tightening environmental regulations can lead to the following effects: (1) decrease of revenues and marginal benefits from innovation and (2) increase of production costs. These effects are caused, for example, by pollution taxes and emission standards (Qiu et al., 2018). These statements were consistent with the traditional view of Palmer, Simpson, and Bradford, who claimed that environmental protection behavior will lead to increased additional costs and weaken a firm's competitiveness (Zhao & Sun, 2016). However, in the 1990s, Porter and van der Linde argued that pollution could be seen as a kind of resource waste, whereas firms' reuse of this waste could lead to improvements in those firms' competitiveness. This paradigm is known as the Porter hypothesis (Porter, 1991; Porter & Van der Linde, 1995). In general, the Porter hypothesis claims that regulatory stringency triggers innovation and consequently allows firms to achieve the dual benefits of environment protection and enhanced firm performance (Hojnik & Ruzzier, 2017).

The Porter hypothesis can be primarily divided into the so-called *weak* and *strong* versions. The weak version of the Porter hypothesis claims that properly designed environmental regulations can spur firm innovations through research and development (Jaffe & Palmer, 1997). The strong version states that properly designed regulations may help increase firm performance in terms of competitiveness and profits (Huiban et al., 2018). In addition, there is also a *narrow* version of the Porter hypothesis, which posits that flexible regulatory policies can give firms greater incentives to innovate and are therefore better than prescriptive forms of regulation, e.g., technology-based standards (Lanoie et al., 2011; Ambec et al., 2013).

The above paradigm inspired a number of researchers and practitioners and therefore launched a variety of additional experiments to verify Porter's hypothesis, which faced a direct response from economists for several reasons. According to Rexhäuser and Rammer (2014), the Porter hypothesis does not offer any comparison of the costs and benefits of environmental regulation. In addition, there are other relevant arguments, such as the very vague theoretical foundations of the Porter hypothesis, which omit a number of other factors influencing firms' (environmental) behavior, and the still unanswered question of whether state regulation is necessary at all. Therefore, Porter's hypothesis also "attracted" several opponents and critics (Palmer et al., 1995). Palmer et al. (1995) criticized Porter's and van der Linde's view and identified two problematic issues, including the roles of both public and private sectors. From the perspective of the public sector, they questioned its ambition to avoid or mitigate the consequences of market failure. In addition, they lacked the inclusion of governmental costs to regulate firms' behavior. By contrast, from the firm perspective, Palmer et al. (1995) argued that state regulations can hamper the emergence of new (profitable) innovations. Hilliard (2004) supports Palmer's critical view and contradicts the statement that environmental regulation can increase national competitiveness by stimulating firm innovation. He states that Porter's hypothesis does not provide a basis that would lead to fundamental changes in the theory of the firm. Rather, it explains the importance of national factors in ensuring the success of firms in various industries.

In order to clearly capture the findings so far, Table 1 offers an overview of selected studies focused on the validity of Porter's hypothesis.

Summing up, in line with the views of Ramanathan et al. (2017), we can state that the findings of prior research on the validity of the Porter hypothesis are still mixed. On the one hand, for example, according to Hu et al. (2020), some studies (see Nesta et al., 2014) examined the positive influence of environmental regulations on innovation performance (as expressed by patents) and R&D expenditures. On the other hand, different studies (see Ambec et al., 2013; Li, 2019) point out that environmental regulatory pressures force firms to invest factors originally used for productive activities into nonproductive activities that reduce pollution emissions, thereby pushing out R&D expenditures.

Despite the above facts, the critics of Porter's hypothesis do not doubt the validity of the hypothesis (in some cases), only challenge its generality (Qiu et al., 2018). Following Huiban et al. (2018: 87), "a difficulty that arises when addressing such a hypothesis is clarifying its interpretation, as the Porter hypothesis is not a hypothesis in a statistical sense but it represents a general idea illustrated with real life examples and, at least in its original formulation, lacked an underlying theory (...) the controversy lies in the strong version, which argues that in many cases this innovation more than offsets the regulatory costs, ultimately enhancing firms' competitiveness and economic performances." Therefore, on the one hand, we can state that quite strong empirical evidence exists of the linkage between environmental regulations and firm innovations, expressed as the weak version of the Porter hypothesis.

Table 1 An overview of selected studies on the validity of the Porter hypothesis

Authors	Research sample	Key findings
Jaffe et al. (1995)	US manufacturing	• The effects of environmental regulation on competitiveness are small and difficult to detect
Simpson and Bradford (1996)	Theoretical strategic trade model	• Strengthening regulation may result in a shift of profits from foreign to domestic firms only in very special cases
Hart (2004)	Innovative environmental new growth model driven by researchers striving for monopoly profits	• Environmental tax may lead to an increase in the rate of production growth • Environmental sales tax may not only provide incentives to reduce pollution but also shift profits from old vintages to new—it could raise incentives to come up with newer (cleaner) vintages
Lanoie et al. (2011)	Approximately 4200 facilities in seven OECD countries (Canada, France, Germany, Hungary, Japan, Norway, the United States) in early 2003	• Strong support for the weak version of Porter hypothesis • Qualified support for the narrow version of Porter hypothesis • No support for the strong version of Porter hypothesis
Böhringer et al. (2012)	German manufacturing sectors between 1996 and 2002	• Environmental regulation could help stimulate environmental investment to become compatible with the pursuit of production growth • There are no positive production impacts induced by environmental or energy expenditures
Nesta et al. (2014)	OECD countries since the late 1970s	• Environmental policies (renewable energy policies) are more effective in fostering green innovation in countries with liberalized energy markets • Environmental policies are crucial only in the generation of high-quality green patents • Competition enhances the generation of low-quality green patents
Rexhäuser and Rammer (2014)	German part of Community Innovation Survey 2008	• Rejection of the strong version of Porter hypothesis • Validity of Porter hypothesis depends on the type of environmental innovation • Innovations that do not improve firms' resource efficiency do not provide positive returns to profitability • Innovations which lead to the increase of firms' resource efficiency have a positive influence on

(continued)

Table 1 (continued)

Authors	Research sample	Key findings
		profitability—relevant for both regulation-induced and voluntary innovations
Ford et al. (2014)	Australian oil and gas industry	• Environmental regulations are related to product and service innovations • Environmental regulations are related to innovations that are both new to the industry and new to the firm (i.e., novel)
D'Agostino (2015)	Review of the empirical evidence	• Multinational firms lack the capabilities to respond creatively to environmental regulation in advanced countries • Multinational firms are attracted to countries with lax environmental regulation
Zhao and Sun (2016)	Chinese pollution-intensive corporations, 2007–2012	• Environmental regulation has significant positive effects on a corporation's innovation • Environmental regulation has negative (insignificant) effects on a corporation's competitiveness • Regional differences arise when verifying the validity of Porter's hypothesis
Ramanathan et al. (2017)	Nine case studies of UK and Chinese firms, based on interviews conducted between 2009 and 2014	• Inflexible environmental regulations force firms to be reactive and adversely affect financial performance • Firms (depending on their resources and capabilities) that adopt either (1) a more dynamic approach to respond to environmental regulations innovatively or (2) a proactive approach to manage their environmental performance are generally better able to reap the private benefits of sustainability
Stoever and Weche (2018)	German business statistics covering manufacturing firms with 20 or more employees for the years 2005–2010	• Environmental regulation (water withdrawal regulation) does not affect firms' overall competitiveness • Analyzed environmental regulation does not qualify as an appropriate policy tool for fostering green growth
Bitat (2018)	German firm-based panel data collected by the Centre for European Economic Research in Mannheim	• Conventional regulatory tools (legally binding instruments) do not trigger innovative behavior at the firm level • A firm's perception of environmental regulations matters more than the type of policy instrument

(continued)

Table 1 (continued)

Authors	Research sample	Key findings
		• Only long-term objectives and market incentives are positively associated with eco-innovation
Zefeng et al. (2018)	Panel data for 1999–2016 from China	• Multidimensional regulations (economic, social, and environmental) have a significant positive impact on regional development • Environmental regulation is more likely to play a role through the channel of multidimensional regulations—therefore, it could be expected that dimensional regulations (economic and social) also have a significant effect in the Porter hypothesis • Environmental regulation may increase costs and therefore have potential inhibitory effects on economic development
Qiu et al. (2018)	Theoretical model of monopolistic competition	• Porter hypothesis holds for high-capability firms, not for low-capability firms • Heterogeneous responses exist in innovation investment • Average industry productivity increases • Strict environmental regulations can encourage firm entry and exit—they could help improve the composition of firms in the regulated industry
Li (2019)	China's industrial sectors from 2005 to 2015	• Environmental regulations have a significant negative impact on technical cooperation and technical introduction • Environmental regulations have no impact on in-house R&D • Environmental regulations have a significant negative impact on technical cooperation, technical introduction, and in-house R&D in high-polluting sectors • Environmental regulations have no significant impact on technical cooperation, technical introduction, or in-house R&D in low-polluting sectors
Javeed et al. (2021)	Pakistani manufacturing firms for 2008 to 2018	• Environmental regulations and corporate social responsibility are positively associated with firm innovation (because of the interactive role of CEO power)

(continued)

Table 1 (continued)

Authors	Research sample	Key findings
		• There exists a positive connection of environmental regulations, corporate social responsibility, and firm innovation with CEO power and ownership concentration—the role of both male and female CEOs for firm innovative practices was confirmed

On the other hand, differing effects are observed for the linkage between environmental regulations and firm competitiveness, expressed as the strong version of the Porter hypothesis (D'Agostino, 2015).

2.2 *Environmental Awareness, Regulations, and Firm Performance*

Although there is no generally accepted consensus among authors on the validity of the Porter hypothesis, we can say that government policy represents an important motivator, which can spur firms' environmental awareness, leading to the creation and diffusion of environmental innovation. As argued by Karakaya et al. (2015: 1092), "policy might have the capacity to create expectations for an environmental innovation and provide clear signals to potential adopters and industrial actors regarding the future attractiveness of it (...) policy makers can foster specific regulations to guide suppliers and adopters to choose an innovation or a specific design, which in return may increase the economical relative advantage of the innovation." By contrast, firms are not pushed only by increasing restrictive environmental regulations on their environmental behavior; there are other forces that raise environmental awareness in firms. Prior literature identified various triggers of firms' environmental awareness and eco-innovation. In this subsection, we therefore present a simplified overview of triggers (determinants) of firms' environmental awareness and eco-innovation. The following sections, in accordance with the purposes of this chapter, focus only on environmental regulations. Table 2 shows the division of triggers of firms' environmental awareness and eco-innovation, according to Rennings (2000) and Horbach et al. (2013).

In line with the above statements about the validity of the Porter hypothesis and the influence of environmental regulations on firms' R&D, innovation, and competitiveness, the question arises whether it "pays to be green" for firms. Referring to Bönte and Dienes (2013) and Doluca et al. (2018), environmental awareness can play an important role in the achievement of sustainable development, but also in the performance of firms, since better environmental performance for a firm can translate into economic success. According to Ramanathan et al. (2017), another key issue is how firms choose to respond to the above triggers of environmental awareness,

Table 2 Different groups of triggers of firms' environmental awareness and eco-innovation

Authors	Group	Examples
Rennings (2000)	Technology push	• Product quality • Material efficiency • Energy efficiency
	Regulatory push	• Existing environmental law • Standards • Expected regulation
	Market pull	• Competition • New markets • Labor costs • Customer demand
Horbach et al. (2013)	Supply-side determinants	• Cost savings • Productivity improvements • Industrial relationships
	Demand-side determinants	• Environmental consciousness • Consumers' preferences for environmentally friendly products
	Regulation and policy determinants	• Implementation and institutionalization of environmental policy instruments • Regulatory design—stringency, flexibility, time frame, anticipation of future environmental regulations

which require them to improve their environmental performance. Considering environmental regulations, Ramanathan et al. (2017: 81) state that "if regulations are flexible enough firms can choose to adopt a dynamic approach to improving environmental performance by: (1) redesigning polluting production processes; (2) the adoption of environmental management practices such as energy conservation and waste management; and (3) strategically positioning themselves as a leader in environmental protection (...) if the firm takes a dynamic approach, flexible regulations could provide opportunities for innovation, which in turn would improve financial performance."

We can also find support for the argument that for firms it pays to be green in studies like of Pereira-Moliner et al. (2012) and Kim (2015). Pereira-Moliner et al. (2012) proved that firm environmental awareness, in the form of environmental management, positively influences several dimensions of firm performance (financial performance, market success, and stakeholder satisfaction) for the sample of Spanish hotels. In addition, Kim (2015) found a positive relation between firms' environmental awareness (behavior) and performance (expressed by sales) by using a sample of 1032 observations divided into specific groups according to the Carbon Disclosure Project Global 500 report and the Dow Jones Sustainability Index between 2008 and 2009.

However, there are also studies that obtained mixed results. For example, Long et al. (2017), by using a sample of 182 Chinese firms in 2014, tested the influence of firms' environmental innovation intentions and environmental innovation behavior on their environmental and economic performance and found several significant differences. Decomposing environmental innovation behavior into product design, raw materials, production processes, and waste treatment, the authors showed, on the one hand, that waste treatment and raw materials do not significantly influence economic performance. On the other hand, they also showed that product design influences firms' economic performance. In addition, Wu et al. (2020) analyzed the effects of firms' corporate environmental responsibility (CER) strength and concern on their innovation performance for the case of Chinese firms listed on the Shenzhen stock exchange between 2006 and 2015. From one perspective, the authors confirmed that CER strength, as expressed by (1) the development of products or services with environmental benefits, (2) the use of clean or renewable energy or resources, (3) pollution prevention, (4) recycling, and (5) an environmental management system, positively influences firms' innovation performance. By contrast, however, their results also showed that CER concern, as expressed by (1) regulation problems, (2) substantial emissions, and (3) energy and resource consumption, negatively influences firms' innovation performance.

Summing up, there is a clear evidence that firms' environmental awareness (triggered by various factors) can positively influence firms' eco-innovation and "non-environmental" performance. Moreover, it can lead to several additional benefits, such as greater resource productivity, a better image and better relationships with (local) stakeholders, higher employee satisfaction, and other benefits (Hojnik & Ruzzier, 2017). However, the question remains what effects environmental awareness will have in CEE countries.

2.3 The Case of Central and Eastern Europe: The Definition of the Research Questions

As we stated above, the countries of Central and Eastern Europe are often perceived as having low environmental awareness, fewer efforts to address environmental issues, and not fully mature awareness of the advantages of eco-innovation (Wysocki, 2021). Support for this statement can be also found in the study by Silajdzic and Mehic (2017), which analyzed the influence of environmental stringency on trade flows among transition economies in Europe. The authors showed that the least regulated CEE countries tend to specialize in pollution-intensive industries and aim to profit from differences in compliance with environmental regulations among EU countries. Such behavior, together with lower innovation performance, makes CEE countries a specific group of "catching-up" countries that are associated with insufficiently developed forms of engagement with environmental matters and with specific histories and sociopolitical developments in their interaction with Western influences (Jehlička & Jacobsson, 2021). Therefore, a commonly held view sees countries of the so-called Eastern bloc as responsible for environmental degradation and pollution across the region, in comparison with

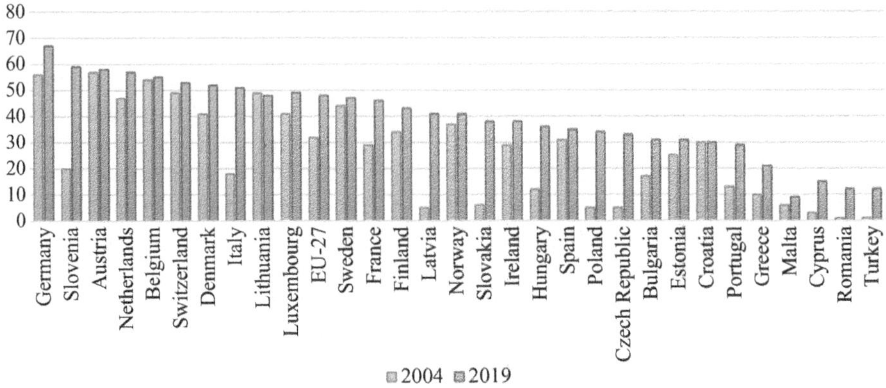

Fig. 1 Municipal waste recycling rates in Europe by country (in %). Source: European Environmental Agency (2022)

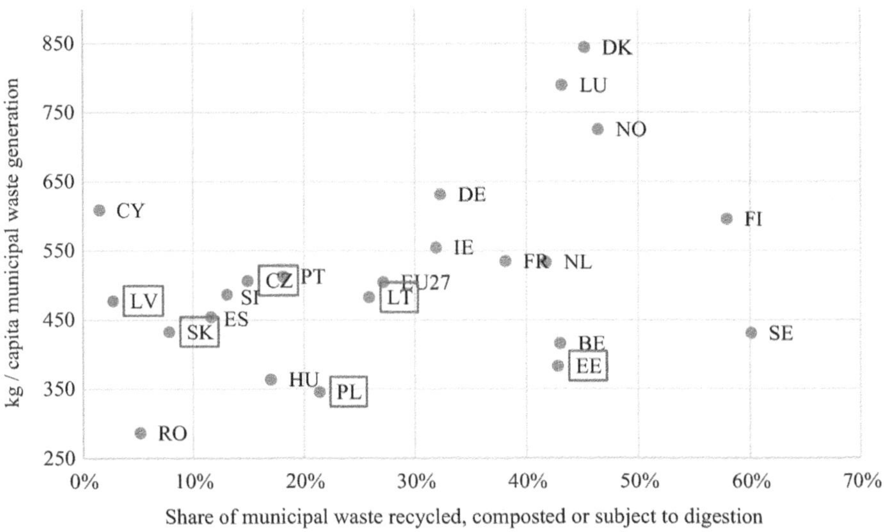

Fig. 2 Country performance in waste generation and recycling in 2020. Legend: The CEE countries analyzed in this chapter are in red squares. Source: Eurostat database

Western, advanced countries, which promoted industrial restructuring in a move toward a resource-saving, energy-efficient society (Tokunaga, 2016).

To have a better picture about lower environmental awareness in CEE countries, we present selected examples that show lower values regarding environmental behavior in CEE countries, compared to their Western and Northern neighbors—please see Figs. 1, 2, and 3. Figure 1 shows municipal waste recycling rates in Europe since 2004, whereas we can see that CEE countries reached lower values compared to their Western and Northern neighbors, especially in 2004. The values

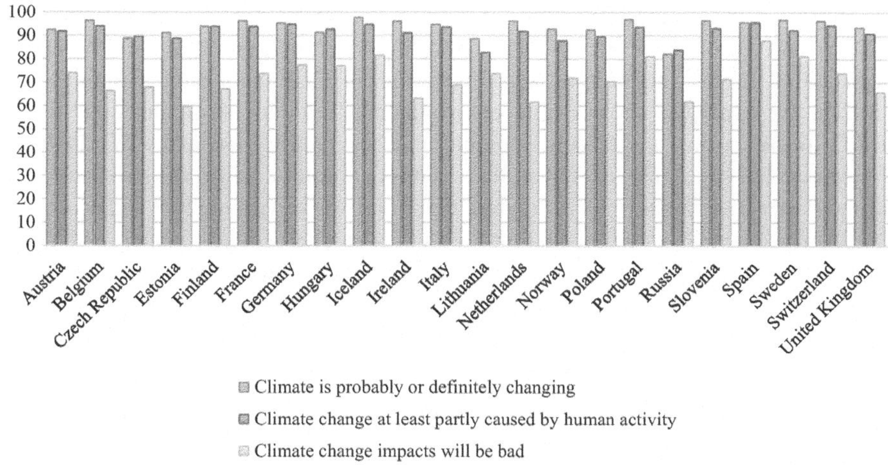

Fig. 3 Countries' beliefs in the reality, causes, and impacts of climate change in 2016–2017 (in %). Source: European Social Survey (2022)

increased significantly between the years 2004 and 2019; however, it is still possible to observe that most of the countries of "Eastern bloc" are lagging behind.

Next, Fig. 2 shows country performance in waste generation and recycling in 2020, whereas we can see that five out of six selected CEE countries (marked with a red square), that we analyze in the next section, reached lower values than the EU-27 average in share of municipal waste recycled, composted, or subject to digestion.

Lower values (in the order of percentages) of the perception of selected issues related to climate change in CEE countries can also be observed in Fig. 3, providing public attitudes to climate change.

For these and other reasons, over the years, it has become widely accepted that Western environmentalism is a benchmark for CEE countries; the knowledge generated in Western contexts has been considered as universally valid, while knowledge from CEE countries was limited to their area only (Jehlička & Jacobsson, 2021). However, current research has shown that procedures and models applied in Western countries are not as "universally valid" for CEE countries. For example, Lewandowska (2016) compared the drivers of eco-innovation in selected CEE countries (Bulgaria, Czech Republic, and Romania) with those in Germany by using the Community Innovation Survey 2006–2008. The results confirmed the assumption that environmental regulations represent a strong determinant of eco-innovation in CEE countries and that financial incentives in this field are far less important. By contrast, the breadth of eco-instruments and regulations was much more varied and important in Germany. Biscione et al. (2021) also confirmed the important role of public policies and incentives (in comparison with demand-pull factors) in the creation of environmental innovation for transition CEE countries in the period 2012–2014. The analyzed CEE countries were Bulgaria, Croatia, Estonia, Hungary, Latvia, Lithuania, Romania, and Slovakia.

However, the results of prior research are not consistent. Despite Lewandowska (2016) and Biscione et al. (2021) confirming the importance of environmental regulations in the creation of eco-innovation in CEE countries, Hojnik et al. (2022), for example, showed that environmental regulations do not spur firms' R&D in the case of the Czech Republic and Slovenia by using the World Bank Enterprise Survey 2019. Moreover, these authors did not confirm the strong version of the Porter hypothesis, which suggests that improved environmental and economic performance can be accomplished through well-designed environmental regulations that induce R&D. Moreover, according to Prause and Olaniyi (2017), environmental regulations can push some sectors into strategic traps due to growing portfolio change requirements and increased restructuring costs—as shown in the example of the Estonian oil shale industry.

Other previous studies have also examined the impact of state regulations and policies on firms' innovation and performance, but more comprehensive research on the validity of the Porter hypothesis in Central and Eastern Europe is (to the best of our knowledge) lacking. For example, Arendt and Grabowski (2019) analyzed the effectiveness of innovation policies in the sample of 820 Polish SMEs. The authors refuted its importance and pointed out that SMEs in less-developed Polish regions rely more on in-house capabilities than on regional innovative potential during the process of creating different types of innovations.

Inconsistency in the prior findings and the limited research on this topic in CEE countries create a research gap that we want to help fill. Therefore, in this chapter, to ensure and increase the generalizability of our results, we create an aggregated data set of selected countries from Central and Eastern Europe (the Czech Republic, Slovakia, Poland, Estonia, Latvia, and Lithuania). More specifically, we focus on countries with several common features: for example, all the selected countries joined the European Union in 2004, and their economic performance does not reach the level of the older member states in Western Europe (Prokop et al., 2021a). Moreover, these countries are below the average performance of EU Member States on environmental innovations, as measured by the summary of the European Commission's Eco-Innovation Index (see https://ec.europa.eu/environment/ecoap/indicators/index_en).

For the purposes of this chapter, given the above arguments, we define three research questions as follows:

RQ_1: Can we confirm the weak and strong versions of the Porter hypothesis in selected CEE countries?

RQ_2: Do environmental regulations affect the environmental awareness of firms in selected CEE countries?

RQ_3: How does environmental awareness affect firms' R&D, innovation, and performance in selected CEE countries?

3 Research Data, Model, and Methods

For the purposes of this chapter, we used an international firm-level survey with a representative sample of an economy's private sector—the World Bank's Enterprise Survey (WBES) 2019. WBES includes various topics focused on the business environment (e.g., performance measures, access to finance, infrastructure, and others) and includes the Green Module (GM), which concentrates on the green economy. GM provides us with information on environment-related issues, management and the environment, environmental policy and regulation, and the environmental impact of the establishment. In total, we analyzed 2139 firms from six countries from Central and Eastern Europe, namely, the Czech Republic, Slovakia, Poland, Estonia, Latvia, and Lithuania. Compared to previous research, in order to ensure a better possibility of generalizing the results, we created an aggregated data set including all firms from the analyzed countries.

Consistent with prior studies (e.g., Su & Cheng, 2019; Hojnik et al., 2022), our research methodology includes two steps: First, we applied confirmatory factor analysis to evaluate the proposed model's (see Fig. 4) reliability and validity. In the second step, we used a variance-based structural equation modeling technique—partial least squares structural equation modeling (PLS-SEM)—to investigate the path analysis of the potential factors and to answer the research questions defined above. The main advantages of PLS-SEM are its robustness against distributional assumptions, high degree of statistical power, and effectively process constructs with small number of items (Hair et al., 2019). PLS-SEM leans on multiple regression analysis and provides scores for the latent variables that could be measured by one or more of the manifest variables (see Table 3).

The general formula of the PLS-SEM model can be expressed as follows:

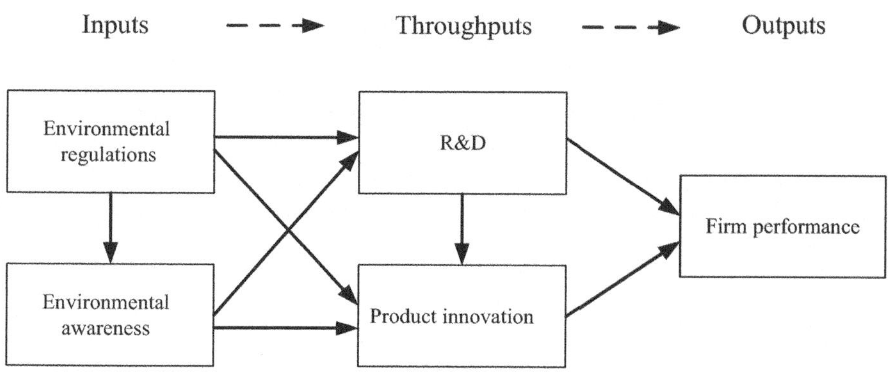

Fig. 4 Proposed research model

Table 3 Description of variables

Latent variable	Manifest variable	Description
Environmental regulation (ER)	Energy tax or levy (ER1)	In the fiscal year, was this establishment subject to an energy tax or levy?
	Energy performance standard (ER2)	In the fiscal year, was this establishment subject to an energy performance standard in its operations?
Environmental awareness (EA)	Environmental strategic objectives (EA1)	In the fiscal year, did this firm have strategic objectives that mentioned environmental or climate change issues?
	Environmental manager (EA2)	In the fiscal year, did this establishment have a manager responsible for environmental and climate change issues?
	Monitoring energy consumption (EA3)	Over the last 3 years, did this establishment monitor its energy consumption?
	Targets for energy consumption (EA4)	Over the last 3 years, did this establishment have targets for energy consumption?
	Energy efficiency measures (EA5)	Over the last 3 years, did this establishment adopt any measures to enhance energy efficiency?
Research and development (R&D)	External knowledge (RD1)	Over the last 3 years, did this establishment spend on the acquisition of external knowledge? This includes the purchase or licensing of patents and non-patented inventions, know-how, and other types of knowledge from other businesses or organizations
	Internal R&D (RD2)	Over the last 3 years, did this establishment spend on research and development activities within the establishment?
	External R&D (RD3)	Over the last 3 years, did this establishment spend on research and development activities contracted with other companies?
	Employees training (RD4)	Over the fiscal year, did this establishment have formal training programs for its permanent, full-time employees?
Control Var. (CV)	Competition (CV1)	In the fiscal year, which of the following was the main market in which this establishment sold its main product? (1, national; 0, international)
	Firm size (CV2)	Binary variable, where 1 = SME and 0 = large firm
Output variable		
Firm performance		The year-on-year sales growth variable

Environmental awareness is expressed in this study as a combination of indicators EA1–EA5, based on the factor analysis

$$\mathrm{EV} = \beta_0^{(k)} + \sum \beta_i^{(k)} z_i + v_k, \tag{1}$$

where:

EV is the explained variable.

$\beta_0^{(k)}$ is the constant term.

$\beta_i^{(k)}$ is the regression coefficient.
z_i is the latent variable.
v_k is the residual term.

In order to ensure and maximize the explanatory power of our models, referring to Ringle et al. (2015), we applied SmartPLS 3 and a consistent PLS bootstrapping technique.

To test the defined research questions, we created a research model, which is presented in Fig. 4. The model consists of inputs that form environmental regulations and environmental awareness. Firm R&D and product innovation (whether or not the firm introduced new or improved products or services) form the throughputs of our model. Firm performance, expressed by the year-on-year sales growth variable (consistent with Kim, 2015), forms the output of our model. In this chapter, we have created two research models that differ by proxy variables for environmental regulations. Energy tax or levy (ER1) is used as a proxy for environmental regulations in Model 1, and energy performance standard (ER2) is used as a proxy for environmental regulations in Model 2. All manifest variables are described in Table 3. Consistent with prior research, we also used two control variables—competition (Sheikh, 2018), expressing whether firms' main market is national or international, and firm size (Arranz et al., 2020), which determines whether the firm is an SME or a large firm.

4 Results

4.1 *Confirmatory Factor Analysis*

In this step, we conducted several tests to ensure the model's validity. Table 4 shows the results on the variance inflation factor (VIF), outer loadings, Cronbach's alpha, rho alpha, and composite reliability. These tests were used to test the collinearity of the formative indicators and the reliability and validity of the constructs. Please note that Model 1 and Model 2 differ in the use of variables that serve as proxies for environmental regulations; therefore, the CFA values for both models are similar.

VIF values were lower than the recommended value of 5 (consistent with Hair et al., 2011). Therefore, we can state that the VIF results implied that collinearity was not a concern in our data. The values of the outer loadings testing the convergent validity were higher than the 0.4 threshold level (Wong, 2016). The values of Cronbach's alpha ($\geq$0.5), rho alpha, and composite reliability ($\geq$0.6) were higher than the recommended values (Durdyev et al., 2018; Hair et al., 2019), except for control variables. However, the values for the control variables were higher than 0.4, which could be expected as their threshold (Abusafiya & Suliman, 2017).

Table 5 shows the tests of model fit or, more specifically, the standardized root mean squared residual (SRMR) criterion (lower than the threshold of 0.08, Gelaye

Table 4 Model reliability and validity

		Model 1—energy tax or levy					Model 2—energy performance standard				
	Var.	VIF	OL	CA	RA	CR	VIF	OL	CA	RA	CR
EA	EA1	1.386	0.625	0.698	0.702	0.695	1.386	0.626	0.698	0.702	0.695
	EA2	1.329	0.505				1.329	0.504			
	EA3	1.368	0.618				1.368	0.614			
	EA4	1.391	0.573				1.391	0.575			
	EA5	1.394	0.471				1.394	0.473			
R&D	RD1	1.137	0.444	0.615	0.624	0.618	1.137	0.448	0.615	0.625	0.619
	RD2	1.431	0.588				1.431	0.593			
	RD3	1.389	0.547				1.389	0.550			
	RD4	1.090	0.563				1.090	0.555			
CV	CV1	1.091	0.540	0.449	0.449	0.449	1.091	0.547	0.449	0.449	0.449
	CV2	1.091	0.536				1.091	0.529			

EA environmental awareness, *R&D* research and development, *CV* control variables, *VIF* variance inflation factor, *OL* outer loading, *CA* Cronbach's alpha, *RA* rho alpha, *CR* composite reliability

Table 5 Models fit

	Model 1	Model 2
SRMR	0.048	0.048
NFI	0.805	0.806

SRMR standardized root mean squared residual, *NFI* normed fit index

Table 6 Discriminant validity—heterotrait-monotrait (HTMT)

	Model 1—Energy tax or levy					Model 2—Energy performance standard				
	CV	EA	ER	INN	R&D	CV	EA	ER	INN	R&D
CV										
EA	0.550					0.550				
ER	0.144	0.048				0.088	0.031			
INN	0.159	0.266	0.037			0.159	0.266	0.041		
R&D	0.613	0.648	0.120	0.407		0.613	0.648	0.106	0.407	
FP	0.030	0.040	0.015	0.040	0.057	0.030	0.040	0.000	0.040	0.057

CV control variables, *EA* environmental awareness, *ER* environmental regulations, *INN* product innovation, *R&D* research and development, *FP* firm performance

et al., 2015) and normed fit index (the value is ≥ 0.80, which suggests a good fit for the model, Saeed & Kasim, 2020). The indicators show a good fit for the model.

We also confirmed the discriminant validity of the model (Table 6) by using the heterotrait-monotrait (HTMT) ratio of the correlation values. The values were lower than the recommended value of 0.85 (Franke & Sarstedt, 2019).

Overall, we can conclude that the construct validity of the presented models is acceptable.

4.2 Experimental Results

4.2.1 Model 1: The Effects of Energy Tax or Levy on Firms' Environmental Awareness, R&D, Product Innovation, and Performance

Figure 5 and Table 7 show the results of the PLS-SEM analyses for Model 1 using energy tax or levy as proxies for environmental regulations.

The results show that environmental regulations (energy tax or levy) significantly influence firms' environmental awareness and R&D. We can see that the influence of environmental regulations on firms' R&D is higher when it is translated through the environmental awareness of the firms. By contrast, there are no significant effects of environmental regulations on firms' innovation and/or performance.

Firms' environmental awareness—which, as we have shown above, is affected by environmental regulations—significantly stimulates research and development of firms in CEE countries (more than does environmental regulation). From one perspective, our results also show that environmental awareness does not directly affect firms' product innovation. From a second perspective, however, we can say that there are significant indirect effects of environmental awareness on firms' product innovation, through R&D.

Interestingly, both environmental regulation and environmental awareness negatively affect firms' performance (through R&D and innovation). However, only in the case of environmental awareness are these effects significant.

Our results also show that firms' R&D represents an important determinant of product innovation in CEE countries. By contrast, we did not reveal positive significant effects of firms' R&D and product innovation on firms' performance. The only significant result in this case shows the negative impact of R&D on firms' performance through product innovation (see Table 7).

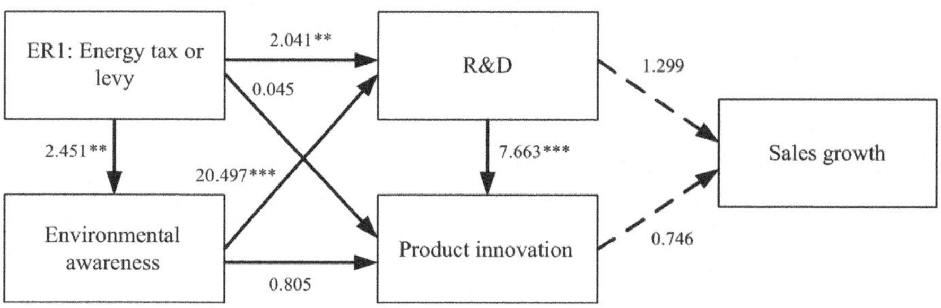

Fig. 5 M1: results for the structural equation models—the direct effects. Solid lines indicate positive effects, dashed lines indicate negative effects, the values on the arrows represent t-statistics; *** significant at $p < 0.01$; ** significant at $p < 0.05$; * significant at $p < 0.10$; the control variable is not shown on this figure (see Table 7)

Table 7 Bootstrapping test on path coefficients (including total indirect effects)—Model 1

	Original sample	Sample mean	Standard deviation	T-stat.	P-Val.
ER1: Energy tax or levy → environmental awareness	0.051	0.052	0.021	2.451	0.015**
ER1: Energy tax or levy → product innovation	0.001	0.004	0.030	0.045	0.964
ER1: Energy tax or levy → R&D	0.086	0.097	0.042	2.041	0.042**
ER1: Energy tax or levy→ R&D → product innovation	0.061	0.069	0.038	1.598	0.111
ER1: Energy tax or levy → environmental awareness → R&D	0.119	0.130	0.044	2.728	0.007***
ER1: Energy tax or levy → R&D → product innovation → firm performance	−0.007	−0.008	0.005	1.381	0.168
Environmental awareness → product innovation	0.042	0.039	0.052	0.805	0.421
Environmental awareness → R&D	0.646	0.644	0.032	20.497	0.000***
Environmental awareness → R&D → product innovation	0.350	0.351	0.045	7.829	0.000***
Environmental awareness → R&D → product innovation → firm performance	−0.040	−0.041	0.022	1.802	0.072*
R&D → product innovation	0.478	0.483	0.062	7.663	0.000***
R&D → firm performance	−0.051	−0.053	0.039	1.299	0.194
R&D → product innovation → firm performance	−0.060	−0.062	0.034	1.784	0.075*
Product innovation → firm performance	−0.020	−0.017	0.027	0.746	0.456
Controls → product innovation	0.159	0.163	0.061	2.588	0.010**
Controls → firm performance	−0.007	−0.006	0.042	0.174	0.862

ER, environmental regulation, *** significant at $p < 0.01$, ** significant at $p < 0.05$, * significant at $p < 0.10$

4.2.2 Model 2: The Effects of Energy Performance Standards on Firms' Environmental Awareness, R&D, Product Innovation, and Performance

Figure 6 and Table 8 show the results of the PLS-SEM analyses for Model 2 using energy performance standard as a proxy for environmental regulations. Similar to the previous environmental regulations, expressed by energy tax or levy, environmental regulations in the form of energy performance standards significantly influence firms' environmental awareness in CEE countries. However, these effects are lower than those of energy tax or levy. This type of environmental regulation also significantly affects firms' R&D, and in addition, it can stimulate the creation of firms' product innovations (indirectly, through R&D).

Regarding environmental awareness, the results of both models confirm the impact of firm environmental awareness on R&D and subsequently (through

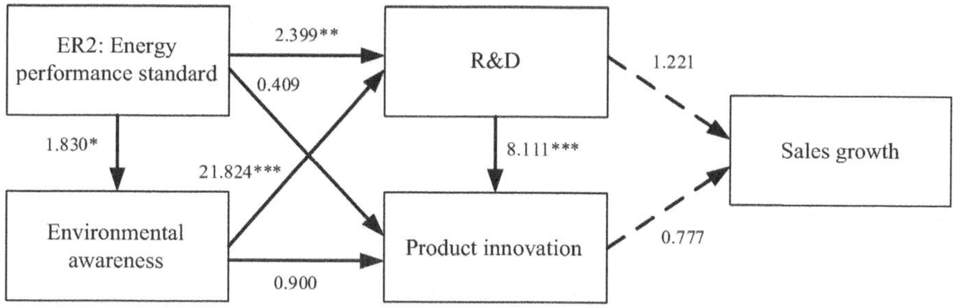

Fig. 6 M2: results for the structural equation models—the direct effects. Solid lines indicate positive effects, dashed lines indicate negative effects, the values on the arrows represent t-statistics; *** significant at $p < 0.01$; ** significant at $p < 0.05$; * significant at $p < 0.10$; the control variable is not shown on this figure (see Table 8)

R&D) on product innovation. The effects of environmental awareness on firms' performance (through R&D and product innovation) in CEE countries are negative.

As in the previous case, firms' R&D represents an important determinant of product innovation among firms in CEE countries.

5 Discussion and Implications

The main contribution of this chapter is that it links the theory and research on (1) the validity of Porter's hypothesis, (2) "catching-up CEE countries," and (3) firms' environmental awareness. Our results confirm the importance of environmental regulations in countries that have been historically criticized for their low environmental awareness and have been identified as higher polluters in Europe than Western and/or Nordic countries. Moreover, we provide insight into the importance of building environmental awareness in CEE countries, because, as we show in Sect. 4, this can also significantly stimulate firms' R&D and "non-environmental" product innovation.

For the first research question—*Can we confirm the weak and strong versions of the Porter hypothesis in selected CEE countries?*—the results are mixed. We showed that both proxies for environmental regulations (energy tax or levy and energy performance standard) act in a similar way regarding firms' R&D in CEE countries. However, Model 2, operating with an energy performance standard, as compared to Model 1, also showed that environmental regulations can influence firms' product innovation—however, in this case, the mediating role of firms' R&D is crucial. These somewhat mixed results allow us to confirm the weak version of Porter's hypothesis, which claims that environmental regulations can spur firm innovations through research and development (Jaffe & Palmer, 1997) only partially. Considering the strong version of the Porter hypothesis, which posits that environmental regulations can also lead to increased firm performance in terms of

Table 8 Bootstrapping test on path coefficients (including total indirect effects)—Model 2

	Original sample	Sample mean	Standard deviation	*T*-stat.	*P*-Val.
ER2: Energy performance standard → Environmental awareness	0.030	0.039	0.017	1.830	0.068*
ER2: Energy performance standard → Product innovation	0.009	0.005	0.023	0.409	0.682
ER2: Energy performance standard → R&D	0.073	0.085	0.030	2.399	0.017**
ER2: Energy performance standard → R&D → Product innovation	0.055	0.059	0.017	3.171	0.002***
ER2: Energy performance standard → Environmental awareness → R&D	0.092	0.110	0.039	2.397	0.017**
ER2: Energy performance standard → R&D → Product innovation → Firm performance	−0.006	−0.006	0.004	1.280	0.201
Environmental awareness → Product innovation	0.044	0.043	0.049	0.900	0.368
Environmental awareness → R&D	0.646	0.645	0.030	21.824	0.000***
Environmental awareness → R&D → Product innovation	0.351	0.351	0.044	7.987	0.000***
Environmental awareness → R&D → Product innovation → Firm performance	−0.040	−0.037	0.024	1.677	0.094*
R&D → Product innovation	0.475	0.476	0.059	8.111	0.000***
R&D → Firm performance	−0.050	−0.046	0.041	1.221	0.223
R&D → Product innovation → Firm performance	−0.060	−0.057	0.037	1.632	0.103
Product innovation → Firm performance	−0.020	−0.022	0.026	0.777	0.438
Controls → Product innovation	0.158	0.157	0.057	2.764	0.006***
Controls → Firm performance	−0.004	0.000	0.045	0.091	0.927

ER, environmental regulation, *** significant at $p < 0.01$, ** significant at $p < 0.05$, * significant at $p < 0.10$

competitiveness and profits (Huiban et al., 2018), our results are consistent and do not confirm it for the case of the selected CEE countries.

The above statements are in line with D'Agostino (2015), who states that there is stronger evidence in the literature for the weak version of the Porter hypothesis but that support for the strong version of the Porter hypothesis is rarer. Referring to Huiban et al. (2018), we can agree with the statement that the controversy lies primarily in the strong version of the Porter hypothesis, which generally posits that innovation more than offsets the regulatory costs and can lead to enhanced competitiveness and economic performance for firms. This has not been confirmed in the selected countries of Central and Eastern Europe, where we can see that environmental regulations trigger actions on the parts of firms but that these actions do not subsequently lead to the growth of the firms' performance, as expressed by sales growth (further discussed below).

For the second research question of this chapter—*Do environmental regulations affect the environmental awareness of firms in selected CEE countries?*—our findings are consistent. Both proxies for environmental regulations trigger firms' environmental awareness in the selected CEE countries. These effects are higher in the case of energy tax or levy. The results corroborate findings from prior research conducted by Biscione et al. (2021). The authors focused on selected transition countries (Estonia, Latvia, Lithuania, and Slovakia) and pointed out the importance of public authorities triggering firms' environmental behavior, as expressed by the creation of eco-innovation. The authors proved these statements by using the Community Innovation Survey 2012–2014. Moreover, our results are also in line with Wysocki (2021), who confirmed that environmental regulations significantly influence firms' environmental initiatives and awareness on a sample of 342 manufacturing enterprises operating in Poland. Wysocki (2021: 15) states "The lack of conviction that innovative green initiatives are economically advantageous is also reflected in the responses related to the potential benefits of their implementation, where political and legal advantages are mentioned in the first place (...) they shape the way companies operate on the market, but they are an expression of the state environmental activity and not the initiatives launched by the companies themselves, which have in a way to comply with environmental requirements out of necessity. This proves a certain carelessness of SMEs with regard to environmental protection and not a fully mature awareness of the advantages of eco-innovation."

If we consider the third research question of this chapter—*How does environmental awareness affect firms' R&D, innovation, and performance in selected CEE countries?*—we can conclude that firms' environmental awareness in CEE countries very strongly affects R&D, which also acts as a mediator of the effects of environmental awareness on firms' product innovation. By contrast, although firms' environmental awareness significantly affects R&D, as well as product innovation (indirectly), firms' performance (as expressed by sales growth) is subsequently negatively affected. These findings contradict the results of Pereira-Moliner et al. (2012) and Kim (2015), who proved that firm environmental awareness can significantly affect several dimensions of firm performance, including sales. However, in the case of CEE countries, it is important to realize that these countries not only have faced environmental awareness problems but also lower innovation performance. This could be caused, among other things, by the abovementioned problems (innovation barriers), such as low absorption capacity, a low level of social capital, insufficient infrastructure, or the inability to effectively draw on national and European funds (Prokop et al., 2021b). Increased environmental awareness may therefore represent completely new activities in the firms, which would require a longer time horizon for positive changes to take effect. Moreover, referring to the above-presented statements of Wysocki (2021), many firms in CEE countries still see state regulation as the main motivation for their environmental behavior and are not aware of the opportunities they could have due to increased environmental perception; therefore, they cannot achieve such performances compared to their Western or Northern neighbors.

Our results also pointed out an important role of firms' R&D in the process of creating product innovations. These results confirmed the previous findings of Prokop et al. (2021a), which state that catching-up CEE countries depend on internal knowledge sources and R&D. The authors also showed that firms in CEE countries still face the challenge of absorbing knowledge and technology from external partners and stated that the benefits of the diffusion of external sources (e.g., foreign technology) can only be delivered if a modern infrastructure, functioning innovation (eco) systems, and governance structures are present in the home countries.

The above findings lead us to several practical implications. First, we recommend continuous support for building and developing internal R&D capacities in the firms of Central and Eastern Europe, accompanied by the provision of access to (and building of) quality infrastructure and technical backgrounds. Next, to ensure firms can also absorb external knowledge and thus achieve the effects of knowledge sharing and spillover between cooperating partners, it is also important to support the creation (and/or increase) of the absorption capacity of firms and their employees (Bönte & Dienes, 2013) and the building of internal and external social capital. Referring to Mazzucchelli et al. (2021), firms' managers should motivate employees to establish and nurture not only internal interpersonal relationships and social ties but also relationships with their external (foreign) partners. For these purposes, they should find ways to increase trust between colleagues and endeavor to share vision and business goals with their employees. This could subsequently lead to increased firm performance and competitiveness.

Another implication concerns the search for examples of good practices, especially in countries that have more experience in applying sustainable business models and in the creation of eco-innovation and are generally recognized as having a high level of environmental awareness. For example, the countries belonging to the performance group of eco-innovation leaders, according to the European Commission's Eco-Innovation Index, are Luxembourg, Finland, Austria, Denmark, Sweden, Germany, France, Spain, and the Netherlands. Although the business models applied in Western Europe may not always work fully in CEE countries, these models should still serve as a source of inspiration for firms in these countries. However, it is necessary to understand both failures and laggards in Western Europe, as well as success stories in CEE countries (to avoid the application of working models in countries where the initial conditions are completely different). Therefore, it is necessary for firms, and also for policy-makers, to learn from these successes and failures and understand "what works" while considering the context and finding context-sensitive ways to disseminate effective solutions across CEE countries. For these purposes, it is crucial to identify country-specific benchmarks and the abovementioned examples of good practices.

Policy-makers should also realize that environmental regulations should not be understood as a scarecrow for firms in CEE countries, but should rather encourage firms to be willing and motivated to act in accordance with the principles of the circular economy. It is therefore necessary to raise higher environmental awareness, both in firms and in society as a whole. Firms and their managers must realize that environmental behavior can also affect their "non-environmental" innovation and performance and consequently result in additional benefits. These could include, for example, increasing resource productivity and employee satisfaction, improving production efficiency, achieving cost savings through the reduction of environmental impact, or creating a better company image (Gadenne et al., 2009; Hojnik & Ruzzier, 2017). Policy-makers should therefore force firms to implement such changes as these will lead to a better understanding of the costs and benefits of their sustainable actions, for example, through government campaigns, promotional advertising, and environmental education interventions within supplier chains (Gadenne et al., 2009).

6 Conclusions

There is no doubt that growing environmental pollution calls for action from all stakeholders in society. However, the role of public policy-makers in this case is debatable, as previous research has shown that state intervention, for example, in the form of environmental regulations, does not always lead to positive effects. This chapter therefore focused on the impact of environmental regulations on firms' R&D, product innovation, and performance. More specifically, we tested the validity of the weak and strong versions of the Porter hypothesis in the case of selected countries from Central and Eastern Europe. In addition, as these countries are often seen as having lower innovation performance and lower perceptions of environmental issues, especially compared to Western European countries, we have also tested the effects of firms' environmental awareness on their R&D, product innovation, and performance. Moreover, we analyzed the influence of environmental regulations on firms' environmental awareness in these CEE countries. The obtained results subsequently allowed us to define practical recommendations for both firms and policy-makers in CEE countries.

We are aware that our analyses have their limitations. In the last part of this chapter, therefore, we present the limitations of our research and recommendations for future research. The first limitation can be seen in the fact that some constructs in our models are relatively small (including also control variables), and it would therefore be appropriate to include more explanatory variables in future research, for example, variables that capture other environmental regulations. This leads us to another recommendation, which is to focus on the narrow version of the Porter hypothesis. A mix of research models is also recommended for future research, such as regression analyses that can help better capture the effects of individual variables. The focus of this study on CEE countries can also be seen as a limitation of our research. Therefore, future research should include the countries of Western Europe to better compare individual countries and then suggest more concrete practical implications. The aggregate data set is another limitation of this study; however, as mentioned above, in this study, we aimed for a greater ability to generalize our results (compared to previous research). As part of future research, we also propose studies on the triggers of environmental awareness in CEE countries or the inclusion of specific firm characteristics in research models—for example, the age of the company, the experience of top managers, gender diversity, and other factors.

Acknowledgment This work was supported by a grant provided by the scientific research project of the Czech Science Foundation [grant No. 20-03037S].

References

Abusafiya, H. A., & Suliman, S. A. M. (2017). Causes and effects of cost overrun on construction project in Bahrain: Part 2 (PLS-SEM path modelling). *Modern Applied Science, 11*(7), 28–37.
Ambec, S., & Barla, P. (2002). A theoretical foundation of the Porter hypothesis. *Economics Letters, 75*(3), 355–360.
Ambec, S., Cohen, M. A., Elgie, S., & Lanoie, P. (2013). The Porter hypothesis at 20: Can environmental regulation enhance innovation and competitiveness? *Review of Environmental Economics and Policy, 7*(1), 2–22.
Arendt, L., & Grabowski, W. (2019). The role of firm-level factors and regional innovation capabilities for polish SMEs. *Journal of Entrepreneurship, Management and Innovation, 15*(3), 11–45.
Arranz, N., Arroyabe, M., Li, J., & Fernandez de Arroyabe, J. C. (2020). Innovation as a driver of eco-innovation in the firm: An approach from the dynamic capabilities theory. *Business Strategy and the Environment, 29*(3), 1494–1503.
Biscione, A., Caruso, R., & de Felice, A. (2021). Environmental innovation in European transition countries. *Applied Economics, 53*(5), 521–535.
Bitat, A. (2018). Environmental regulation and eco-innovation: The Porter hypothesis refined. *Eurasian Business Review, 8*(3), 299–321.
Böhringer, C., Moslener, U., Oberndorfer, U., & Ziegler, A. (2012). Clean and productive? Empirical evidence from the German manufacturing industry. *Research Policy, 41*(2), 442–451.
Bönte, W., & Dienes, C. (2013). Environmental innovations and strategies for the development of new production technologies: Empirical evidence from Europe. *Business Strategy and the Environment, 22*(8), 501–516.
Bu, M., Qiao, Z., & Liu, B. (2020). Voluntary environmental regulation and firm innovation in China. *Economic Modelling, 89*, 10–18.
Chizaryfard, A., Trucco, P., & Nuur, C. (2021). The transformation to a circular economy: Framing an evolutionary view. *Journal of Evolutionary Economics, 31*(2), 475–504.
D'Agostino, L. M. (2015). How MNEs respond to environmental regulation: Integrating the Porter hypothesis and the pollution haven hypothesis. *Economia Politica, 32*(2), 245–269.
Doluca, H., Wagner, M., & Block, J. (2018). Sustainability and environmental behaviour in family firms: A longitudinal analysis of environment-related activities, innovation and performance. *Business Strategy and the Environment, 27*(1), 152–172.
Durdyev, S., Ismail, S., Ihtiyar, A., Bakar, N. F. S. A., & Darko, A. (2018). A partial least squares structural equation modeling (PLS-SEM) of barriers to sustainable construction in Malaysia. *Journal of Cleaner Production, 204*, 564–572.
European Commission. European innovation scoreboard; internal market, industry, entrepreneurship and SMEs. Accessed November 26, 2021., from https://ec.europa.eu/growth/industry/policy/innovation/scoreboards_en
European Environmental Agency. Waste recycling in Europe. Accessed April 1, 2022., from https://www.eea.europa.eu/ims/waste-recycling-in-europe

European Social Survey. European Attitudes to Climate Change and Energy: Topline Results from Round 8 of the European Social Survey. Accessed April 1, 2022., from https://www.europeansocialsurvey.org/docs/findings/ESS8_toplines_issue_9_climatechange.pdf

Ford, J. A., Steen, J., & Verreynne, M. L. (2014). How environmental regulations affect innovation in the Australian oil and gas industry: Going beyond the Porter hypothesis. *Journal of Cleaner Production, 84*, 204–213.

Franke, G., & Sarstedt, M. (2019). Heuristics versus statistics in discriminant validity testing: A comparison of four procedures. *Internet Research, 29*(3), 430–447. https://doi.org/10.1108/IntR-12-2017-0515

Gadenne, D. L., Kennedy, J., & McKeiver, C. (2009). An empirical study of environmental awareness and practices in SMEs. *Journal of Business Ethics, 84*(1), 45–63.

Geissdoerfer, M., Savaget, P., Bocken, N. M., & Hultink, E. J. (2017). The circular economy–a new sustainability paradigm? *Journal of Cleaner Production, 143*, 757–768.

Gelaye, B., Tadesse, M. G., Lohsoonthorn, V., Lertmeharit, S., Pensuksan, W. C., Sanchez, S. E., Lemma, S., Berhane, Y., Vélez, J. C., Barbosa, C., Anderade, A., & Williams, M. A. (2015). Psychometric properties and factor structure of the general health questionnaire as a screening tool for anxiety and depressive symptoms in a multi-national study of young adults. *Journal of Affective Disorders, 187*, 197–202.

Hair, J. F., Ringle, C. M., & Sarstedt, M. (2011). PLS-SEM: Indeed a silver bullet. *Journal of Marketing Theory and Practice, 19*(2), 139–152.

Hair, J. F., Risher, J. J., Sarstedt, M., & Ringle, C. M. (2019). When to use and how to report the results of PLS-SEM. *European Business Review, 31*(1), 2–24.

Hart, R. (2004). Growth, environment and innovation—A model with production vintages and environmentally oriented research. *Journal of Environmental Economics and Management, 48*(3), 1078–1098.

Hilliard, R. (2004). Conflicting views: Neoclassical, porterian, and evolutionary approaches to the analysis of the environmental regulation of industrial activity. *Journal of Economic Issues, 38*(2), 509–517.

Hojnik, J., & Ruzzier, M. (2017). Does it pay to be eco? The mediating role of competitive benefits and the effect of ISO14001. *European Management Journal, 35*(5), 581–594.

Hojnik, J., Prokop, V., & Stejskal, J. (2022). R&D as bridge to sustainable development? Case of Czech Republic and Slovenia. *Corporate Social Responsibility and Environmental Management, 29*(1), 146–160.

Horbach, J. (2016). Empirical determinants of eco-innovation in European countries using the community innovation survey. *Environmental Innovation and Societal Transitions, 19*, 1–14.

Horbach, J., Oltra, V., & Belin, J. (2013). Determinants and specificities of eco-innovations compared to other innovations—An econometric analysis for the French and German industry based on the community innovation survey. *Industry and Innovation, 20*(6), 523–543.

Hu, J., Pan, X., & Huang, Q. (2020). Quantity or quality? The impacts of environmental regulation on firms' innovation–quasi-natural experiment based on China's carbon emissions trading pilot. *Technological Forecasting and Social Change, 158*, 120122.

Huiban, J. P., Mastromarco, C., Musolesi, A., & Simioni, M. (2018). Reconciling the Porter hypothesis with the traditional paradigm about environmental regulation: A nonparametric approach. *Journal of Productivity Analysis, 50*(3), 85–100.

Jaffe, A. B., & Palmer, K. (1997). Environmental regulation and innovation: a panel data study. *Review of Economics and Statistics, 79*(4), 610–619.

Jaffe, A. B., Peterson, S. R., Portney, P. R., & Stavins, R. N. (1995). Environmental regulation and the competitiveness of US manufacturing: What does the evidence tell us? *Journal of Economic Literature, 33*(1), 132–163.

Javeed, S. A., Latief, R., Jiang, T., San Ong, T., & Tang, Y. (2021). How environmental regulations and corporate social responsibility affect the firm innovation with the moderating role of chief executive officer (CEO) power and ownership concentration? *Journal of Cleaner Production, 308*, 127212.

Jehlička, P., & Jacobsson, K. (2021). The importance of recognizing difference: Rethinking central and east European environmentalism. *Political Geography, 87*, 102379.

Karakaya, E., Hidalgo, A., & Nuur, C. (2015). Motivators for adoption of photovoltaic systems at grid parity: A case study from southern Germany. *Renewable and Sustainable Energy Reviews, 43*, 1090–1098.

Kim, Y. (2015). Environmental, sustainable behaviors and innovation of firms during the financial crisis. *Business Strategy and the Environment, 24*(1), 58–72.

Korhonen, J., Nuur, C., Feldmann, A., & Birkie, S. E. (2018). Circular economy as an essentially contested concept. *Journal of Cleaner Production, 175*, 544–552.

Kotkova Striteska, M., & Prokop, V. (2020). Dynamic innovation strategy model in practice of innovation leaders and followers in CEE countries—A prerequisite for building innovative ecosystems. *Sustainability, 12*(9), 3918.

Lanoie, P., Laurent-Lucchetti, J., Johnstone, N., & Ambec, S. (2011). Environmental policy, innovation and performance: New insights on the Porter hypothesis. *Journal of Economics & Management Strategy, 20*(3), 803–842.

Lewandowska, M. S. (2016). Do government policies foster environmental performance of enterprises from CEE region? *Comparative Economic Research. Central and Eastern Europe, 19*(3), 45–67.

Li, C. (2019). How does environmental regulation affect different approaches of technical progress?—Evidence from China's industrial sectors from 2005 to 2015. *Journal of Cleaner Production, 209*, 572–580.

Long, X., Chen, Y., Du, J., Oh, K., Han, I., & Yan, J. (2017). The effect of environmental innovation behavior on economic and environmental performance of 182 Chinese firms. *Journal of Cleaner Production, 166*, 1274–1282.

Mazzucchelli, A., Chierici, R., Tortora, D., & Fontana, S. (2021). Innovation capability in geographically dispersed R&D teams: The role of social capital and IT support. *Journal of Business Research, 128*, 742–751.

Nesta, L., Vona, F., & Nicolli, F. (2014). Environmental policies, competition and innovation in renewable energy. *Journal of Environmental Economics and Management, 67*(3), 396–411.

Palmer, K., Oates, W. E., & Portney, P. R. (1995). Tightening environmental standards: The benefit-cost or the no-cost paradigm? *Journal of Economic Perspectives, 9*(4), 119–132.

Pereira-Moliner, J., Claver-Cortés, E., Molina-Azorín, J. F., & Tarí, J. J. (2012). Quality management, environmental management and firm performance: Direct and mediating effects in the hotel industry. *Journal of Cleaner Production, 37*, 82–92.

Porter, M. E. (1991). America s green strategy. *Scientific America, 264*, 168.

Porter, M. E., & Van der Linde, C. (1995). Toward a new conception of the environment-competitiveness relationship. *Journal of Economic Perspectives, 9*(4), 97–118.

Prause, G., & Olaniyi, E. O. (2017). The impact of environmental regulations on regional development in Eastern Estonia. In *New challenges of economic and business development–2017: Digital Economy* (pp. 461–472).

Prokop, V., Stejskal, J., Klimova, V., & Zitek, V. (2021a). The role of foreign technologies and R&D in innovation processes within catching-up CEE countries. *PLoS One, 16*(4), e0250307.

Prokop, V., Striteska, M. K., & Stejskal, J. (2021b). Fostering Czech firms' innovation performance through efficient cooperation. *Oeconomia Copernicana, 12*(3), 671–700.

Qiu, L. D., Zhou, M., & Wei, X. (2018). Regulation, innovation, and firm selection: The porter hypothesis under monopolistic competition. *Journal of Environmental Economics and Management, 92*, 638–658.

Ramanathan, R., He, Q., Black, A., Ghobadian, A., & Gallear, D. (2017). Environmental regulations, innovation and firm performance: A revisit of the Porter hypothesis. *Journal of Cleaner Production, 155*, 79–92.

Rennings, K. (2000). Redefining innovation—Eco-innovation research and the contribution from ecological economics. *Ecological Economics, 32*(2), 319–332.

Rexhäuser, S., & Rammer, C. (2014). Environmental innovations and firm profitability: Unmasking the Porter hypothesis. *Environmental and Resource Economics, 57*(1), 145–167.
Ringle, C. M., Wende, S., & Becker, J.-M. (2015). *SmartPLS*. SmartPLS.
Saeed, A. F., & Kasim, N. (2020). Influencing factors governance (IFG) model for United Arab Emirates disaster management. *Psychology and Education Journal, 57*(9), 5464–5471.
Sheikh, S. (2018). The impact of market competition on the relation between CEO power and firm innovation. *Journal of Multinational Financial Management, 44*, 36–50.
Silajdzic, S., & Mehic, E. (2017). The impact of environmental taxes on competitive performance of pollution-intensive industries among transition economies: Evidence from panel analysis. In *Financial environment and business development* (pp. 155–169). Springer.
Simpson, R. D., & Bradford, R. L., III. (1996). Taxing variable cost: Environmental regulation as industrial policy. *Journal of Environmental Economics and Management, 30*(3), 282–300.
Stoever, J., & Weche, J. P. (2018). Environmental regulation and sustainable competitiveness: Evaluating the role of firm-level green investments in the context of the Porter hypothesis. *Environmental and Resource Economics, 70*(2), 429–455.
Su, C. H., & Cheng, T. W. (2019). A sustainability innovation experiential learning model for virtual reality chemistry laboratory: An empirical study with PLS-SEM and IPMA. *Sustainability, 11*(4), 1027.
Tokunaga, M. (2016). Regime change and environmental reform: A systematic review of research on Central and Eastern Europe. *Economic Review, 67*(4), 339–353.
Wong, K. K. K. (2016). Mediation analysis, categorical moderation analysis, and higher-order constructs modeling in partial least squares structural equation modeling (PLS-SEM): A B2B example using SmartPLS. *Marketing Bulletin, 26*(1), 1–22.
Wu, W., Liang, Z., & Zhang, Q. (2020). Effects of corporate environmental responsibility strength and concern on innovation performance: The moderating role of firm visibility. *Corporate Social Responsibility and Environmental Management, 27*(3), 1487–1497.
Wysocki, J. (2021). Innovative green initiatives in the manufacturing SME sector in Poland. *Sustainability, 13*(4), 2386.
Zefeng, M., Gang, Z., Xiaorui, X., Yongmin, S., & Junjiao, H. (2018). The extension of the Porter hypothesis: Can the role of environmental regulation on economic development be affected by other dimensional regulations? *Journal of Cleaner Production, 203*, 933–942.
Zhao, X., & Sun, B. (2016). The influence of Chinese environmental regulation on corporation innovation and competitiveness. *Journal of Cleaner Production, 112*, 1528–1536.